Beginning MongoDB Atlas with .NET

Flexible and Scalable Document Data Storage for .NET Developers

Luce Carter

Beginning MongoDB Atlas with .NET: Flexible and Scalable Document Data Storage for .NET Developers

Luce Carter
Manchester, UK

ISBN-13 (pbk): 978-1-4842-9549-6
https://doi.org/10.1007/978-1-4842-9550-2

ISBN-13 (electronic): 978-1-4842-9550-2

Managing Director, Apress Media LLC: Welmoed Spahr
Acquisitions Editor: Jonathan Gennick
Development Editor: Laura Berendson
Editorial Project Manager: Shaul Elson

Cover designed by eStudioCalamar

Cover image by Freepik (www.freepik.com)

Distributed to the book trade worldwide by Springer Science+Business Media LLC, 1 New York Plaza, Suite 4600, New York, NY 10004. Phone 1-800-SPRINGER, fax (201) 348-4505, e-mail orders-ny@springer-sbm.com, or visit www.springeronline.com. Apress Media, LLC is a California LLC and the sole member (owner) is Springer Science + Business Media Finance Inc (SSBM Finance Inc). SSBM Finance Inc is a **Delaware** corporation.

For information on translations, please e-mail booktranslations@springernature.com; for reprint, paperback, or audio rights, please e-mail bookpermissions@springernature.com.

Apress titles may be purchased in bulk for academic, corporate, or promotional use. eBook versions and licenses are also available for most titles. For more information, reference our Print and eBook Bulk Sales web page at http://www.apress.com/bulk-sales.

Any source code or other supplementary material referenced by the author in this book is available to readers on GitHub (https://github.com/Apress). For more detailed information, please visit https://www.apress.com/gp/services/source-code.

If disposing of this product, please recycle the paper

For Jay,

You encouraged me to become an author when the chance arose, never doubting I could do it. You never stopped believing in me or being proud of my achievements. You were one of my biggest cheerleaders and I will never forget that. Sadly, we lost you before I finished this book, but I miss you every day.

Table of Contents

About the Author xi

Acknowledgments xiii

Introduction xv

Part I: Getting Started 1

Chapter 1: Choosing MongoDB 3

Relational vs. NoSQL Databases 3

  What Is a Relational Database? 3

Advantages of a Relational Database 6

  Simplicity 6

  Normalization 6

  Atomicity, Consistency, Isolation, Durability (ACID) 7

  Data Accuracy 8

Disadvantages of a Relational Database 8

  Performance 8

  Scalability 8

  Flexibility 9

  Cost 9

  Complexity 9

What Is a NoSQL Database? 9

  Key-Value Database 10

  Wide-Column Database 13

  Graph Database 14

  Document Database 17

Summary 20

Chapter 2: What Is MongoDB? **21**

The Beginning of MongoDB 21

MongoDB Server (On-Premises) 23

    Enterprise Advanced 23

    Community Edition 23

MongoDB Atlas 24

    Database 24

    Aggregation 27

    Serverless 30

    Data Federation 30

    Data Lake 31

    Search 31

    Atlas CLI 31

Atlas App Services 32

    Triggers 32

    Functions 32

    Authentication 33

    Data API 34

Realm 34

    Device Sync 35

Charts 35

Summary 36

Part II: Setting Up MongoDB **39**

Chapter 3: Creating an Account **41**

Creating Your MongoDB Account 41

Google Account 42

GitHub Account 44

Create a New Account 45

Finalizing Your New Account 48

Summary 50

Chapter 4: Creating Your First Cluster and Loading Sample Dataset 51

Creating Your First Cluster 51

Loading Sample Dataset 56

Summary 59

Chapter 5: Browsing Your Data 61

Atlas UI 61

    Reading Data 61

    Creating or Updating Data 64

    Deleting Data 67

Mongosh 68

    Reading Data 71

    Creating Data 72

    Updating Data 72

    Deleting Data 73

MongoDB Compass 73

    Reading Data 74

    Creating, Updating, or Deleting Data 77

    Aggregation Builder 78

Visual Studio Code Extension 81

    Reading Data 83

    Creating Data 84

    Updating Data 85

    Deleting Data 85

Data API 86

    Reading Data 87

    Creating Data 88

    Updating Data 89

    Deleting Data 89

Drivers 90

GraphQL 91

    Reading Data 95

    Creating, Updating, and Deleting Data 95

Summary 97

Part III: Building a Project 99

Chapter 6: Creating the Application 101

Tooling 102

    NET SDK – CLI 102

    Integrated Development Environment (IDE) 105

Cleaning Up 108

Summary 108

Chapter 7: Adding MongoDB 111

Add the MongoDB NuGet Package 111

Store Connection String 112

    Fetch Your Connection String from Atlas 112

    Add the Connection String to Our Application 114

Create a MongoDB Client 115

Retrieving a List of Databases 116

Final Code 117

Summary 119

Chapter 8: Creating and Interacting with Documents from Code 121

Adding the Model 121

Creating the Service 123

Updating to Use the Service 124

Adding Create Methods to the Service 124

Adding Read Methods to the Service 125

Adding an Update Method to the Service 125

Adding a Delete Method to the Service 126

Creating a Controller 126

Final Code 130

    Program.cs 130

    Game.cs 131

    GamesDatabaseService.cs 132

    GamesController.cs 133

Summary 135

Chapter 9: Testing the Endpoints 137

Creating Documents 138

    Creating One Document 138

    Creating Many Documents 139

Reading Documents 140

    Reading All Documents 140

    Reading One Document 141

Updating a Document 142

Deleting a Document 143

Summary 144

Part IV: Taking It Further 145

Chapter 10: Schema Validation 147

Data Modeling 147

Validating for Required Fields 148

    Handling Invalid Documents 150

Specifying Data Types for Fields 151

Specifying Allowed Values for a Field 152

Applying Validation to Existing Documents 154

    Allowing Invalid Documents on an Ad Hoc Basis 156

    Finding Invalid Documents to Update 156

Schema Anti-Patterns 159

Summary 160

Chapter 11: What Next? **161**

Adding a UI 161

    Using .NET MAUI and Atlas Device SDKs 162

Visualizing Your Data 162

Searching Your Data 163

    Atlas Search 163

    Atlas Vector Search 163

Play with the Sample Data 164

Learn More 164

Summary 165

Index **167**

About the Author

 **Luce Carter** is a Developer Advocate for MongoDB with a passion for sharing knowledge and making technology and code seem less intimidating. She is a Microsoft MVP and an international public speaker, enjoying speaking at conferences and other local meetups to share things she is passionate about. When not at a computer, she can be found playing squash with her local club, swimming, or trying to find interesting new places to walk. Her work to educate developers includes helping them to battle imposter syndrome – one line of code and story at a time.

About the Author

Acknowledgments

Much of this book would not have happened without various people in my life for different reasons.

First, my mum, who has had to listen to me talk about a range of things and be a chatterbox my entire life. You are wonderful! You always checked in on how this book went, even though you don't truly understand what it is about.

Megan, a great friend and my copy editor. The quality of this book is better because of you. Thanks for always reviewing chapters to a great standard and being excited about my writing.

Kayy, for supporting me when I needed a distraction or to free my brain of something on my mind so I could focus on the book.

Rita, for allowing me to take time to write the book when I needed to and providing me with an amazing job and colleagues that make me want to write a book.

Alex and Rick, for tech-reviewing the book at different stages.

Kailie, for believing in me and making me feel like a rockstar for writing a book, convincing me to do it when I doubted myself.

Josh, who invited me to speak at his user group, giving a talk on this topic for the first time, that led to this book existing.

Northern Rebound, the best squash club I could have possibly joined. You all took an interest, gave me a mental break each week, and whose weekly social gave me a place to go early every week to get the book over the line.

Introduction

Welcome to *Beginning MongoDB Atlas with .NET*. A lot of time and love has gone into this book, so I hope you enjoy reading it as much as I enjoyed writing it.

I never imagined I would become a published author, but when the chance came about, I couldn't resist. I have loved helping the community for many years, often public speaking or creating content. This led me to the honor of being recognized as a Microsoft MVP in 2018. Around this time, I discovered that there is a role that is like being an MVP but paid: Developer Advocate.

In June 2021, I was lucky enough to become a Developer Advocate for the first time, working at MongoDB. To help me get up to speed on MongoDB as a developer data platform, I committed to giving a talk on MongoDB Atlas with .NET at the Liverpool .NET user group.

A short while after I delivered this talk, an acquisition editor at Apress reached out after seeing I had delivered the talk and asked if I thought there was enough content in this area for a book. This got me thinking and I started to picture what the structure of the book might look like. Before I knew it, I was putting together a proposal to become the author of this very book.

Luckily for me, Apress accepted my proposal and the wheels were in motion. The proposal acted as a guide to each chapter, so some of the work had been done before I even began. So when I was accepted and could get started, I was able to start quickly.

Like with everything in technology, MongoDB products can move fast. So when you read this, screenshots may be out of date or versions different, so please bear that in mind. There may even be features available now that weren't in existence when I wrote the book!

This book is structured to be both informational and educational. It starts with a walk through of types of databases and the history of MongoDB, including why it was created.

After that, it becomes a hands-on tutorial, showing you how to get started with the C# driver using a Web API project. It starts with deploying your first cluster, creating the project, and hooking it all together. You can follow it chapter by chapter, or even just drop in to a chapter for a reminder of how to do something later on.

PART I

Getting Started

CHAPTER 1

Choosing MongoDB

Technology is famous for moving at lightning speed. Things are forever changing and adapting, and new technologies and frameworks become popular. But there are some features that are often present in the majority of projects and applications. Examples include the *user interface (UI)*, network connectivity, and data storage.

Data is everywhere, whether it is the timetable for the public transport we take, our browsing history on our devices, our medical history, our finances, or something else entirely. This data needs to be stored somewhere so it can be accessed from one or many applications. This is where *databases* come in.

When it comes to choosing a database technology and vendor, there are many choices for your projects. This chapter aims to discuss and compare the different options as well as go further into what exactly *MongoDB* offers and where it fits in.

Relational vs. NoSQL Databases

Once you have decided that you need a database in your project, the first choice you face is what kind of database technology to go with.

Traditionally, the most common type of database has been a *relational*, or *tabular*, *database*. However, as technology has evolved, a competing set of options has surfaced: *NoSQL (not only SQL) databases*.

What Is a Relational Database?

A relational database, or *relational database management system (RDBMS)*, stores data in *tables*. A database can be made up of one or more tables. These tables store related data, and often, data is shared between tables to form relationships between them. This is where the name "relational database" comes from.

© Luce Carter 2024
L. Carter, *Beginning MongoDB Atlas with .NET*, https://doi.org/10.1007/978-1-4842-9550-2_1

Inside these tables, *columns* are used to define the data and its shape, and *rows* are used to hold the records of this data. At least one column will be specified as holding unique values which identifies each row, and this column will become known as the *primary key*. This primary key is then used in other tables to form the relationships, whereby it becomes the *foreign key* in that table.

Let's look at an example of how data might be represented in a relational database to better understand this concept. Imagine we own our own business selling games. This business would have customers, products, orders, and suppliers. Customers would order products, and we would replenish products from our suppliers. We can see an example of how this might look across tables below in Table 1-1 for our customers, Table 1-2 for storing products, and Table 1-3 for orders.

Table 1-1. *Example of how data is represented in a relational database*

CustomerId	Full_Name	Address
1	Joe Bloggs	42 Data Lane, Information Land, SW1 1DB
2	Lisa Smith	18 Apple Way, Fruit Corner, N14 4RD
3	John Doe	27 Word Street, Binary, B12 1DN

Table 1-2. *Example of how the products table might look*

ProductId	Name	Price	Category
1	Monopoly	19.00	Board Game
2	Uno	9.99	Card Game
3	Carcassonne	6.99	Board Game

Table 1-3. *Example of how the orders table might look*

OrderId	ProductId	Quantity	CustomerId
1	1	1	2
2	3	2	1
3	2	1	1

These tables show how the ids for other tables are used to share information between them. For example, the orders table doesn't need to know product names or addresses, but instead, it can use the product and customer ids to look up that information when required, saving on storage space.

You may also hear relational databases referred to as "SQL" (sometimes pronounced "sequel"). This comes from *structured query language* (SQL) which is the language used to build, query, search, or filter one or more tables for the required data. SQL has both ANSI and ISO SQL standards and these have been evolving since 1986, the most recent being SQL:2016.

There are a few main providers when it comes to relational databases: *Oracle SQL*, *MySQL*, *Microsoft's SQL Server*, and *PostgreSQL*. Each of these has their own slight variation of SQL dialect, some with additional extensions. For example, SQL Server from Microsoft is the most different from the standard, using a dialect of SQL called *T-SQL* or Transact-SQL.

MySQL is the most popular open source database product. PostgreSQL is becoming more popular and is another open source database with advanced features. PostgreSQL is probably the best one for newcomers to relational databases today, as it has a free tier as well as a commercial distribution and has the most syntax in common with other variations should you need to use another type of RDBMS.

SQL Server is often used by Enterprises. There is a free community edition should you want to get started for free.

Oracle makes both Oracle SQL and MySQL. However, Oracle SQL is a paid-for product and considered more "commercial" as it is not open source and therefore not as easily changed.

As previously mentioned, there are slight syntax variations in the SQL implementations between the different databases. These differences are only small in basic queries, but once you move into more complex queries, the differences can add up and become quite significant.

The following code snippet shows a simple query in SQL which will fetch all the data for Lisa Smith from the customers table in Table 1-1. This syntax will be valid and generate the same results across all the main RDBMS.

```
Select *
FROM customers
WHERE
Full_name = 'Lisa Smith';
```

Although relational databases are not the focus of this book, it is good to consider the advantages and disadvantages to help better understand why this type of database may or may not suit your needs.

Advantages of a Relational Database

Compared to what came before, the relational database concept brought numerous advantages to the table that were welcomed by developers who had grown weary of doing battle with other database models of the day. The following sections detail some of these advantages that early programmers struggled to achieve and that we take for granted today.

Simplicity

One of the great things about relational databases is their simplicity. Although they are complex in some ways, they have been around for so long that a variety of resources and tools have been developed to make them easier to use. SQL as a query language is also understandable by more than just developers, making it accessible to business analysts and more.

SQL also made it possible to retrieve data from a database without having to write a lot of code.

Normalization

Although the cost of storage is a lot cheaper now, when relational databases were first created, this was not always the case. Normalization can be done to different levels or normal forms and involves eliminating or reducing anomalies, which in turn reduces the amount of storage space required to store the data.

Atomicity, Consistency, Isolation, Durability (ACID)

Relational databases use transactions to carry out operations. *ACID* is a standard that guarantees a level of reliability of these transactions. Each transaction consists of one or more operations. The letters in ACID refer to these four properties:

Atomicity – The principle of atomicity is that if one operation fails, the whole transaction fails, and the database will rollback any changes made before the error occurred.

Consistency – Relational databases often have constraints on the data via implicit or explicit rules. The principle of consistency states that if any operation in a transaction breaks this integrity, the transaction will fail.

Isolation – Isolation is a guarantee that concurrent transactions on a database won't impact any others. This does not mean that operations can't happen in parallel, but a global queue will be used inside the database to ensure that no outcome is affected if the operation will be carried out on the same rows or tables.

Durability – Durability is the promise that a backing store such as a changelog will be used, so in the event of a system failure, the data that was successfully committed will be retained. This means that when the database reports that a transaction has been successfully completed, it has been saved to this backing store for reference in the event of failure.

An example of why ACID is important can be seen in financial institutions. Imagine you are transferring £40,000 to the bank to pay the deposit to buy your first house. If the transaction goes wrong somewhere in the absence of ACID, you might find that the money leaves your account but never arrives in the recipient's account, so you are left out of pocket. Disaster!

The database prevents this by keeping a record of all operations, and if one of them fails, it rolls back the entire transaction.

This also protects you if someone sinister is attempting to steal money from your bank account or use your credit card details. If there is not enough money in the account to make a payment, it will be declined due to constraints applied by the bank.

Data Accuracy

As mentioned earlier, relational databases form relationships between tables by making use of keys. As explained already, the primary key field always has unique values, and this is also true when it is used as the foreign key in another table. This means you can be more confident of a lack of data duplication.

Disadvantages of a Relational Database

Although around for a long time, relational databases have disadvantages, which are some of the reasons why other types of databases have gained popularity. The following sections describe some of these disadvantages.

Performance

With relational databases, data is stored across multiple rows in multiple tables, and more complex queries will require expensive joins. These joins bring the data together from multiple tables, but if there is a lot of data to process across many tables, this can lead to slow performance.

Scalability

Traditionally, relational databases have been created with the intention that they will be run on a single computer or server. This means that if the requirements of that machine are not sufficient – for example, because the database is being accessed more frequently either by users or applications – the only solution is to add additional hardware. This is called *vertical scaling*.

However, there is a ceiling to how much new hardware or better hardware you can add to a machine. Plus, as you add more resources, the cost efficiency will have diminishing returns.

Flexibility

Relational databases use rigid schemas. For example, schemas can be used to define column names, data types, and constraints. Examples of data types include text, number, and date. You can then add constraints such as a format for dates, the maximum number of characters in a text field, or not allowing a price field to be a negative number.

However, this rigidity can be an issue. If a change to the schema is required, the database may have to be taken offline while the changes are applied. This can also lead to some less than desirable workarounds – for example, adding a new column that is only required for certain records, meaning that for all existing records, the value of the new columns might have to be NULL.

Cost

As well as the cost of the hardware if scaling vertically, there are other costs associated with any type of database, including relational databases that should be considered. A lot of the enterprise software for managing a database also comes with licensing costs. There is also the cost of hiring and employing someone to maintain the database.

Complexity

Relational databases can also be complex. Once the size of the dataset increases, the number of tables and rows can increase exponentially. This means that the data requires expensive joins and queries to bring the required data together when generating a report.

This means that queries may become too complex to write and understand, for a non-technical person to use.

What Is a NoSQL Database?

A NoSQL database is used to describe any database technology that isn't subject to the limitations of the tabular format of a traditional relational database. Many different NoSQL databases have been developed over the years, either to overcome disadvantages of relational databases or to address a specific use case.

There are many different types of NoSQL databases. For the rest of this chapter, we will look at examples of some of these different types and their advantages and disadvantages.

Key-Value Database

A key-value, or key-store, database is one of the most basic types of databases.

Data is stored in key/value pairs, like in a JavaScript Object Notation (JSON) object. The key is a unique value used to reference a specific value. The value can be a variety of data types from text or number to more complex objects.

This is similar in concept to a dictionary, hash table, or map object from many programming languages. However, it is instead stored inside a database and managed via the database management system.

What Are the Main Features of a Key-Value Database?

Keys are names or identifiers for the value and can be used to retrieve the values. Often, a key-value database will also allow for the insertion, deletion, and updating of a value via its key, not just retrieving it.

What Are Examples and Use Cases for Key-Value Databases?

Key-value databases are probably more common than you think. For example, the first key-value database I probably interacted with, without even knowing it, is the Windows Registry.

In the Windows Registry – shown in Figure 1-1 – both system and application values are stored using keys and values. You are not advised to change the registry as this can cause unexpected issues, but often, the more technically savvy would do things to make our lives easier or more pleasant when interacting with Windows.

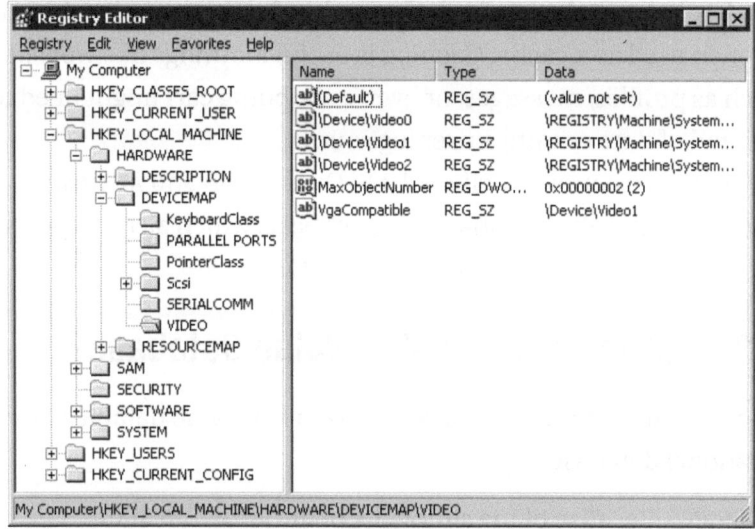

Figure 1-1. *Windows Registry*

Another example of a key-value database is the settings for an application. Often, developers will use a key-value database for settings if there is a large amount to store – for example, in a game. Each setting will have a name, such as "Invert_Y-Axis," for looking or flying a plane in game, and this will have a value assigned.

Table 1-4 shows what a key-value database for a game might look like, if displayed in table format.

Table 1-4. *Example of a game setting key-value database*

Key	Value

Resolution	{
	"width": 1920,
	"height": 1080
	}

Difficulty	"Easy"
Invert_Y-Axis	False
Subtitles	True

Key-value databases are best for simple scenarios where a key and associated value are enough, ideally when there is real-time random access.

There are a few big competitors in the key-value database market. The most widely used is *Redis*, often used as a cache. Caching is useful for things like website sessions where data such as profiles, messages, or even a shopping cart might need to be saved for a certain period of time or until a user logs out.

However, other big examples include *AWS' DynamoDB* and *Memcached*.

Another popular key-value database is Couchbase, which also supports text searches and querying, using a SQL-like language called N1QL.

What Are the Advantages of a Key-Value Database?

The following are some of the advantages to using a key-value database over the more traditional, relational database:

- Efficiency – Because of the simplicity of the structure of the data, there is no need for complex indexing, meaning reading and writing is much faster.

- Simplicity – Because it is just a unique identifying key and associated value, it doesn't require a complex language such as SQL to query the information.

- Scalability – Key-value databases can be easily horizontally scaled, making them flexible to adapting requirements in terms of usage and much more cost-effective than scaling vertically.

What Are the Disadvantages of a Key-Value Database?

As well as the advantages mentioned previously, the following are some of the disadvantages to using a key-value database:

- Limited use cases – Although very useful, because of the simple structure, there is a limit to the use cases for this type of database. Once the requirements get more complex, it can't really be supported by a key-value database.

- No filtering – The data is stored as a blob, so it is not possible to filter on value fields.

- Values can only be updated as a whole – This might seem trivial, but in some scenarios, this inability to update only part of the value can add to complexity or slow performance – for example, if using a key-value database for a shopping basket on an e-commerce website and the user adds an additional item to the basket.

Wide-Column Database

Wide-column databases are like relational databases in that they store data in tables, columns, and rows. However, the columns can span multiple servers or files. The formatting and names don't have to match in each row, either. They are also known as a multi-dimensional key-value store because they store one value potentially against multiple columns.

Examples and Use Cases for Wide-Column Databases

Wide-column databases are great for situations where there is a large amount of data that may not have consistent columns for each row but may be distributed across multiple servers or nodes.

Internet of Things (IoT) sensor data is one example. The various sensors may not all use the same data structure, but it will insert data regularly, causing the size of the database to increase very quickly.

Time series data is another example, such as from finance, where data like stock price over time might be logged at regular intervals to track things such as irregularities.

Apache Cassandra is a common example of a wide-column database.

Advantages of Wide-Column Databases

Here are some examples of the advantages of wide-column databases over relational databases:

- Flexibility – The shape of the data doesn't have to be consistent, allowing for flexibility around the type of data logged, even if it could change. It can handle "big data" such as those generated by constantly logging information.

- Partial updates – A lot of wide-column database systems support the partial update of column value or even adding a new one, since it can simply create a new file to store that column.

- Performance – The lack of a complex structure leads to fast queries.

Disadvantages of Wide-Column Databases

Here are some of the disadvantages of wide-column databases:

- Performance – Because data is distributed across multiple files or servers, although some queries can be incredibly fast, updating can be slow. This is because attributes are stored together as columns rather than rows, so transactions must be carried out across multiple locations. Wide-column databases are better for analytical processing rather than transactional.

- Limited use case – Like with key-value databases, there is a limit to the use cases because of their simplicity.

Graph Database

Graph databases are considered one of the most specialized of the NoSQL databases. They specialize in relationships between entities. This might sound like relational databases, but it is more advanced than that.

What Are the Main Features of a Graph Database?

The following is a list of the main features that differentiate a graph database from other types of databases:

- Nodes – They store information about the things in your database. These are sometimes called vertices.

- Edges – They store information about the relationships or actions between the notes in the database. These are sometimes called relationships.

- Properties – Metadata about a node or edge, also called properties, is stored in a label as a key-value pair.

- Label – This is optional but can be used to apply tags to a group of related nodes. It's a bit like how a table holds related data in a relational database.

What Are Examples and Use Cases of a Graph Database?

Let's now look at an example of how you might store data in a graph database. In this example, we will be looking at how to store data about rail travel.

Each railway station would be represented by a node, and data about the station would be stored in the node's properties. For example, my nearest railway station is Manchester Piccadilly in the UK. This could have data such as station_code: MAN and location: Manchester.

A train journey between two stations would be represented by an edge as it is the relationship between two nodes. There are regular trains between Manchester and London, so we will use that as an example. Figure 1-2 shows that journey. A node represents both stations, Manchester Piccadilly and London Easton. Then, an edge represents the relationship between them, which is the train journey. The properties for the edge could be train_time: 0937 and train_operating_company: AvantiWestCost

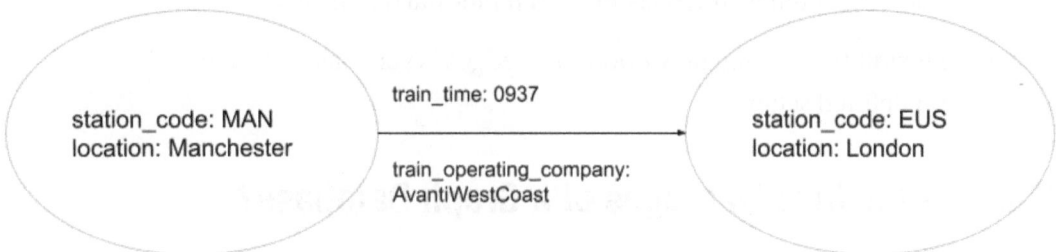

Figure 1-2. *Graphical representation of a train journey*

You may also choose to add more data and relationships to the data. You might add nodes for other related entities such as station shops. You could optionally apply a label like "Station" to the station nodes and "shop" to the shop nodes.

Graph databases are intended for situations where the relationships between nodes are important – for example, in identifying patterns. A huge use case for graph databases is in the world of fraud detection. Fraudsters have gotten clever with how they attempt to hide their criminal activity using various accounts, often based in international banks. Graph databases can help map out the relationships and actions between these accounts.

Netflix uses a graph database for their discovery engine, which helps them recommend TV shows and movies to a user based on their activity and the activity of others – for example, what shows a user often watched next after finishing something, which can help suggest what another user might want to see after finishing the same content.

The most well-known graph database is Neo4j.

What Are the Advantages of a Graph Database?

Here are some of the advantages of a graph database:

- Analytics – As mentioned already, because of the ability to store information not just about an entity but also the relationships between them, graph databases are fantastic for analytics-based scenarios where pattern detection is important.

- Performance – For queries relating to the relationship between entities, graph databases carry these out incredibly fast. The indexes are generated automatically through the relationships, meaning querying them is much faster than relational databases.

- Flexibility – Adding new nodes and edges is very easy. Plus, there is no defined schema.

What Are the Disadvantages of a Graph Database?

The following is a list of some of the disadvantages of a graph database:

- Querying unrelated data – If there is a lack of defined relationships, or not much recorded in the way of relationships, querying becomes difficult in graph databases.

- Query language – There is no clearly defined query language in graph databases, like there is in relational databases with SQL. This means that there is a learning curve when moving between graph database technologies.

- Performance – Graph databases are not good at handling many transactions at once. They are also not intended for database-wide queries, so these can be difficult or slow.

- Scalability – Like relational databases, information is often stored on one server rather than multiple.

Document Database

Now, for my favorite general-purpose database, *document databases*!

What Are the Main Features of a Document Database?

Document databases store data in documents, which are often *JavaScript Object Notation (JSON)*-like structures that support a wide variety of data types, from strings to numbers to dates, raw binary, arrays, and even other objects. Data is often stored in key-value pairs. However, documents can also be more like *XML*.

Documents are the conceptual equivalent of a row/record in relational databases. The difference being that a row must be flat, whereas a document can be hierarchical. Each document typically stores information about one specific entity/object and any related data. Groups of documents are stored in a collection, the equivalent of a table in a relational.

What Are Examples and Use Cases of a Document Database?

Imagine we want to store data about employees in an HR system. With a document database, you can create a document for each user with relevant data such as name, address, and job history. In a relational database, this wouldn't be achieved with one table. You would have a separate table for job history information with columns for title, date started, and date ended. If the employee is still working in one job, there wouldn't be a date ended value, so this would have to be set as NULL, which is considered bad practice.

A document could be created using JSON that stores all this data in one place.

```
{
    "name": "Joanne Bloggs",
    "address": "42 Data Lane, Database City, DB14 8JD",
```

```
    "email": "jo@data.com",
    "phone_number": 07123456789,
    "job_history": [
        {
            "title": "Database Administrator",
            "date_started": "01-06-2021"
        },
        {
            "title": "Database Assistant",
            "date_started": "18-05-2018",
            "date_ended": "24-05-2021"
        }
    ]
}
```

Each employee at the company would have a document, and these would be stored in an employee collection. This would be stored inside a database, potentially with other collections, such as a collection of documents about offices.

One of the great things about document databases is their flexibility. Because they are a general-purpose NoSQL database, they can be used for a variety of use cases and can, in fact, be considered a superset of other database types in some ways.

A document database may have simple key-value-based fields. In the earlier employee example, name, address, phone number, and email were all examples of a single value assigned to a key.

Even relational databases can be more cleverly modeled in a document database using embedded documents and arrays. Data that is accessed together should be stored together for faster retrieval, so by embedding information that is related inside a single document, it only requires one call to fetch a document to get all this information back.

Graph databases can also be similarly recreated using a document database. This is by modeling nodes and/or edges as documents.

Due to the general-purpose nature of a document database, it can be used for many different use cases. Some examples include

- Customer data management

- IoT data

- Time series data

- Mobile apps

- Analytics

- Product catalogs

There are a few big names who have a document database product. This includes Microsoft's *Azure* which has a NoSQL, JSON-like document store option available as part of their database product called CosmosDB, and of course, we have MongoDB. MongoDB can be run both locally on a server using MongDB Community or MongoDB Enterprise, or in the cloud, using *MongoDB Atlas*.

What Are the Advantages of a Document Database?

There are many advantages of using a document database over traditional tabular, relational databases, and the following list gives some of those advantages:

- Readability – Developers are often used to JSON objects from languages such as C++, Python, C#, and many more, meaning reading a document is intuitive and easy because document databases often use a JSON-like structure.

- Flexibility – A document can have many fields, but not every document has to have the same fields, even within a collection. It is easy to add or exclude fields as requirements evolve, and this won't have any impact or require downtime.

- Scalability – Databases can be distributed across multiple servers, allowing for horizontal scaling.

- Resilience – Copies can be run in parallel as a secondary data source and be promoted to primary in the event of an issue.

- Performance – Keeping related data together in one document minimizes the need for expensive joins, meaning queries are much faster. Transactions overall are also faster for things like updates because you are only updating in one place.

What Are the Disadvantages of a Document Database?

There are, however, some disadvantages of document databases. The following lists some of these:

- Duplication – Storing related data together in documents, instead of uniquely in related tables, could duplicate data across documents.

- Foreign keys – This often isn't required based on how the data is modeled, but document databases do not support foreign keys.

So Why Choose MongoDB?

MongoDB is so much more than just a generic document database. It is strongly general-purpose, being able to support a wide variety of use cases. It was also built to be highly available and flexible, intending to offer a great experience but also solve some of the limitations of relational databases. In Chapter 2, we will dive deeper into MongoDB and how its suite of full developer data platform products enables developers to achieve faster, with a lower friction experience.

Summary

We have made it to the end of the first chapter! In this chapter, you have learned the following:

- Relational databases have been around a long time and offer lots of tools and resources.

- However, they can be quite complex, especially as the data increases in size and leads to more and more tables.

- NoSQL comes in different types and is suited for different purposes.

- Document databases, including MongoDB, which is a great general-purpose solution for a variety of scenarios.

In the next chapter, we will take a further look at the history of MongoDB and the variety of products on offer from a cloud database to charts to mobile databases.

CHAPTER 2

What Is MongoDB?

Now that we know more about relational vs. NoSQL databases and have seen the wonders of document databases, it is time to understand more about MongoDB – from the beginning to now, including the different products available for so many different use cases.

In this chapter, we will look through more of why MongoDB was created and go into detail of the breadth of products available as part of the developer data platform, including the database and the additional services available.

The Beginning of MongoDB

MongoDB was founded in 2007 by Dwight Merriman, Eliot Horowitz, and Kevin Ryan. Interestingly, they all worked together at DoubleClick, the Internet advertising company now owned by Google.

While working at DoubleClick, the team created many custom data stores in a bid to work around shortcomings in existing database offerings. The business served around 400,000 ads per second, and they found themselves struggling with scalability and agility. This is what inspired the creation of what we now know as MongoDB!

At first, MongoDB was known as 10gen, Inc, and this is one company that Dwight, Eliot, and Kevin founded after leaving DoubleClick, following the Google acquisition. The aim was to build a *platform as a service (PaaS)* that was centered around the web and built using only open source components. They were also fans of the idea of *horizontal scaling*. They saw a future of the web where vertical scaling alone wasn't enough.

In an increasingly data and information-driven world, the size of datasets is getting infinitely bigger. Horizontal scaling is a solution to growing datasets, spreading the workload across multiple servers. The complexities of managing across multiple servers are taken care of with MongoDB.

© Luce Carter 2024
L. Carter, *Beginning MongoDB Atlas with .NET*, https://doi.org/10.1007/978-1-4842-9550-2_2

In 2008, the MongoDB product was first open-sourced, and it has been at least partially open source ever since.

In 2012, MongoDB University was born and launched the first online MongoDB course. In 2022, MongoDB University was rebranded as MongoDB Learn.

Then in 2013, the company reached another milestone in its history: the rebrand.

As well as the existing support and training services offered, 10gen expanded to also offer an enterprise version that included Kerberos integration for on-premises monitoring services. Monitoring of deployments was also available for free in the cloud as well as a cloud backup and restoration service.

After this is when 10gen became MongoDB and the existing products for monitoring and backup were integrated into MongoDB Management Services (MMS). MMS was available as both an on-premises and cloud product, specializing in the monitoring, automation, and backup of clusters. This has since been rebranded. MMS is now MongoDB Cloud Manager, with the on-premises software called MongoDB Ops Manager.

As the years went on, MongoDB made subtle but significant changes to its offerings, including a marketing update to change the messaging behind MongoDB Enterprise and what was at the time called MMS.

The year 2016 was also the year they introduced Atlas, their cloud-based database as a service.

In 2017, MongoDB 3.6 was released. This introduced Change Streams, a product for always-on, real-time, reactive applications for more resilience, including with retryable writes. Another key feature of 3.6 was the introduction of schema validation, to help enforce data integrity. This is also the year that MongoDB became a public company, listing on NASDAQ.

2018 was the year that a big major version came out: 4.0. This release introduced multi-document ACID transactions. This was significant because it opened up a broader range of use cases and made migrating from legacy databases easier.

MongoDB 4.2 came out in 2019 and this included client-side field-level encryption (CSFLE), meaning the database server only ever deals with encrypted ciphertext.

In 2020, MongoDB 4.4 was released. This included Atlas Online Archive, to automatically archive data from your database to fully managed, queryable object storage. 4.4 also added multi-cloud deployments, adding the ability to distribute data in a single cluster across multiple public clouds simultaneously, or move workloads easily between them.

The year 2021 and the release of 5.0 brought in the first preview (since been released) versions of serverless instances on Atlas for dynamic scaling and cost-based on usage.

We have now "caught up" somewhat in the timeline to start looking at what we have today. The preceding paragraphs are highlights of some of the things introduced in those releases. But a lot of these introduced products and features that are still available now. Atlas is a core part of this book, so for the rest of this chapter, we are going to look at the different products that MongoDB has now, not just Atlas but many others too, and discuss what they are and why they exist.

MongoDB Server (On-Premises)

The first database product provided by MongoDB was installed, managed, and controlled by database administrators, often on servers in data centers. This is still available to this day.

It is more than just a database, however. It is a collection of tools that allow for better security and efficiency of the deployed MongoDB database.

Enterprise Advanced

This is the on-premises database server intended for professional/enterprise use.

You can automate the administration of the database using MongoDB Ops Manager, the management tooling for on-premises deployments. Ops Manager also allows for the backup, restoration, and monitoring of those deployments. Plus, there are advanced access and data security features, including auditing and encryption, to protect the database and meet high security compliance.

It is also possible to connect MongoDB Server to other tools using the MongoDB Connector for BI. This allows read-only SQL support, enabling you to query your data in MongoDB with SQL using tools such as Tableau, Power BI, and even Excel, for better business intelligence visualizations.

Community Edition

The community edition offers the same developer features and flexible, distributed document data model as Enterprise, but without some of the more advanced security and management features including auditing, LDAP and Kerberos auth, Encrypted Storage Engine, In-Memory Storage Engine, and MongoDB Ops Manager.

It is licensed under the Server Side Public License (SSPL).

MongoDB Atlas

Once upon a time, MongoDB Atlas was MongoDB's database as a service in the cloud. But now it has evolved to be far more than that. It is a multi-cloud developer data platform with many powerful features. Let's briefly discover these different features.

Database

The core product at the heart of Atlas is still the document database as a service. This is multi-cloud, allowing a single cluster to be deployed to multiple cloud providers, including Microsoft's Azure, Google's Cloud Platform, and Amazon's AWS. This is great for resilience and data mobility without manual intervention. It even goes a step further and supports multi-region, allowing for data to be stored closer to the users. In conjunction with multi-cloud, it can support scenarios such as data sovereignty requirements in the face of a cloud provider region outage.

Because it is hosted in the cloud, Atlas simplifies deploying, managing, and scaling your database to suit your needs.

As you will learn more about later, we have lots of tools available for interacting with your Atlas clusters in a really easy way.

Replica Sets

One of the key features under the hood of MongoDB databases is replica sets. All Atlas clusters have a minimum of at least one replica set, even with the free tier cluster. In a replica set, three or more nodes all contain the same data. One of these nodes will be considered the primary node, and all read and write operations will target it by default. However, with the use of read preferences, other nodes (called secondary nodes) can be targeted. Figure 2-1 illustrates this.

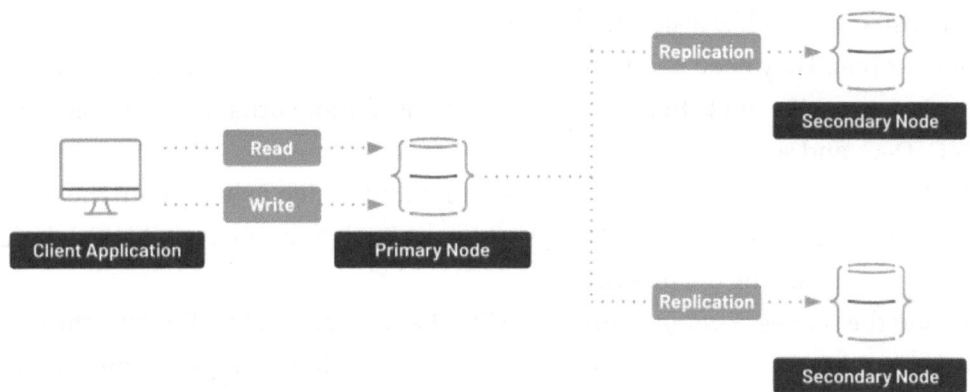

Figure 2-1. *Replication in MongoDB with a primary node and two secondary nodes*

A key feature as well is what is called read and write concerns. Write concern is what is required to receive a write success acknowledgement, requested from the node being written to. The most common default write concern level is majority. This means that a majority of the nodes have reported writing the data. Once this happens, a success acknowledgment is returned. There are also other options, such as only requiring the primary node to complete before being considered successful, primary and secondary node, all nodes, or no acknowledgement required at all.

Read concern gives control over the isolation and consistency of data read from replica sets and also shards (that we will learn about shortly).

Data is kept consistent between nodes, and if the primary node goes down, one of the secondary nodes becomes the primary, and a new secondary node is created in the background to replace it.

This can all be managed via both the Atlas UI and the CLI tools we have available.

This is what helps MongoDB be such a highly available database. It also helps reduce bottlenecks on read operations. However, in an application with many write operations, the changes need to be propagated across all the replica set members, which can lead to issues with performance. This is where sharding comes in.

Sharding

Sharding is the preferred method of horizontal scaling (also known as scaling out). In sharding, data is split across multiple nodes. Rather than the data being consistent across all the nodes like in replication, each shard, consisting of three or more replica nodes, contains a subset of the data. Because there is a subset, this is excellent for use cases with high write frequency as it only needs to apply the changes to the node the data changed is stored in.

You can think of it like aisles in a library, for example. There are too many books to just put in a pile. They need to be spread across bookshelves, in a system that makes it easy to find a specific book. In this case, it is done by author surname, with aisles named like A–C, D–G, and so on.

In MongoDB, a shard key would be used to identify which field to shard the data on. In the library example, this is the author's surname. Each bookshelf would be a shard, with the books being the documents.

Behind the scenes, a *mongos* instance will act as a query router, directing the query to the correct shard. A *config server* then stores configuration settings and metadata, as shown in Figure 2-2.

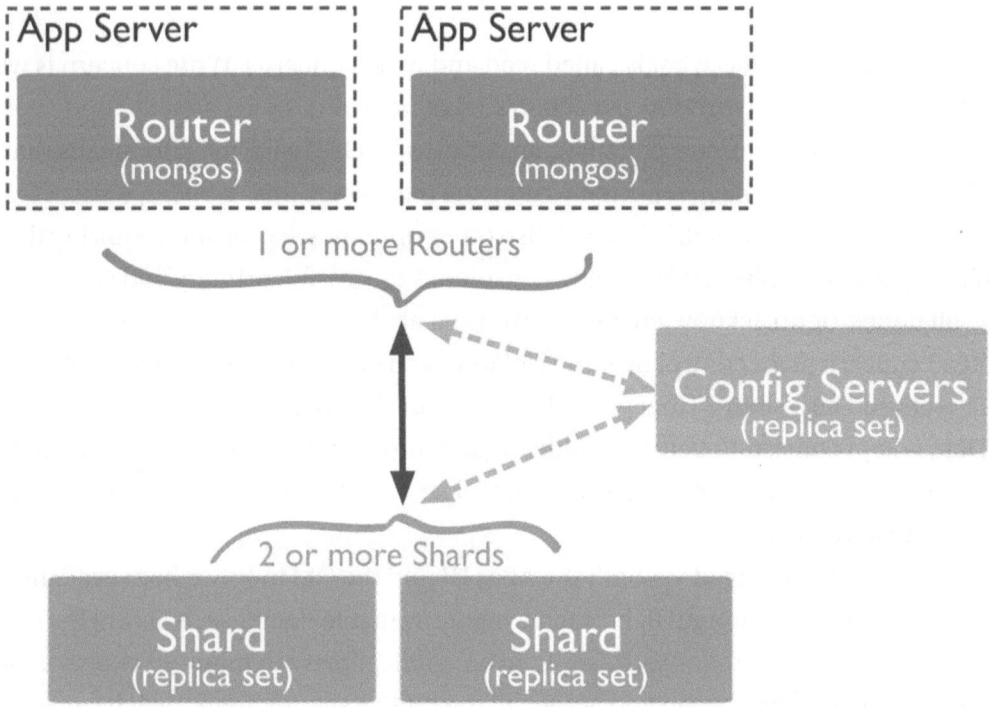

Figure 2-2. *MongoDB sharding*

Sharding can handle both horizontal and vertical scaling. If being scaled vertically, more nodes can be added to each shard. If being scaled horizontally, more shards can be added. The great thing is MongoDB will handle rebalancing the data between the shards, as the scaling occurs.

Sharding can also be broken down into three different strategies depending on the use case:

1. Ranged sharding – Documents are grouped into ranges of shard key values called chunks. By default, a chunk is about 128MB of ranged document data. All documents within that given chunk range are then stored on a single shard. The sharded cluster balancer takes care of splitting and distributing chunks across multiple shards.

2. Hashed sharding – A hash of the shard key (either a single field hashed index or a compound hashed index) is used to distribute the documents. It can help create a uniform distribution of writes across shards. They are great for high-volume write scenarios where documents are most commonly retrieved individually.

3. Zoned sharding – This is a special type of ranged sharing and allows you to associate shards with any number of zones. Zones are associated with one or more clusters. The cluster balancer migrates chunks covered by a zone to only the shards associated with that zone. Sharding can be deployed to MongoDB Atlas in M30 (or higher) tier dedicated Atlas clusters. Sharding in Atlas is easy to configure. In the settings for the cluster, either at creation time or later, sharding can be enabled and you just select the number of required shards.

Aggregation

In MongoDB, the most basic query you can carry out on your data is find(), which is like a select in SQL. The command can take a filter to help match documents, or be called without any parameters to return every document in a collection. You can also use a projection parameter to select which fields in a document you want to return, avoiding sending data back that you don't need.

However, as queries get more advanced and complex, you might find that the basic find() command isn't as flexible as you would like in terms of reshaping, merging, and applying calculations to your data. This is where *aggregation* queries come into play.

Aggregation is a multi-stage pipeline, with each stage transforming documents from the previous stage before returning a final aggregated result. It was designed specifically to improve usability and performance. It does not need to have a 1:1 input:output ratio and can generate new documents and even filter some out. As of MongoDB 4.4, it even supports custom expressions.

You can think of the aggregation pipeline as a factory floor and conveyor belt system, with each section a stage of the aggregation pipeline. The documents as they are live at the start of the conveyor belt and each stage helps select, filter, or transform the output, which then becomes the input for the next stage.

There are several operators that you can perform in your aggregation pipeline. The following is a list of some examples:

- `$multiply`

- `$sum`

- `$range`

- `$or`

- `$switch`

- `$concat`

- `$unwind`

Now let's look at an example.

If we consider our shop example from Chapter 1 and imagine it is stored as a document database, we might want to know the average spend each day. We might have a collection of transactions where each document looks like Figure 2-3.

```
date: 2003-09-09T00:00:00.000+00:00
amount: 7514
transaction_code: "buy"
symbol: "adbe"
price: "19.10728026500741805193683831021189689636 23046875"
total: "143572.1039112657392422534031"
```

Figure 2-3. *Example transaction document*

We can take advantage of a number of available stages and operators, to help us calculate this new average. We can use $set to create a new field, which is the result of applying a $avg operator to the price field. However, you will notice that price is stored as a string, so as part of calculating the average, we can use the $toDecimal operator to convert it into the right data type. We can then round the result to two decimal places with $round to get it in a format that looks like price.

When written in code, this could look similar to the below example, with each stage applying to the previous until we get what we want.

```
[
  {
    '$group': {
      '_id': '$transactions.date',
      'avgSpend': {
        '$avg': {
          '$toDecimal': '$transactions.price'
        }
      }
    }
  }, {
    '$project': {
      'avgSpendPerDay': {
        '$round': [
          '$avgSpend', 2
        ]
      }
    }
  }
]
```

Serverless

One of the new features of MongoDB Atlas introduced in 2021 is *Serverless* instances. It is excellent for when you don't know what your workload will be, or how much traffic the database will get. It can easily scale to adapt to the changes, without you needing to know any of this upfront. This often leads to lower costs, as you only pay for the resources you use. However, if the workload has the likelihood to be predictable, then a committed resource tier, such as our dedicated clusters, may be more cost-effective.

In a traditional database as a service, such as a MongoDB Atlas database, you provision an instance size at creation time, having to predict size and resources required ahead of time – although auto-scaling is available for gradual increases in workload so there is some flexibility. You can reduce the risk of running out of resources with dedicated clusters by planning to scale up before workload spikes, such as with bulk inserts.

However, with Serverless, it is more reactive to change. You create the serverless instance for your data, and the resources will scale to reflect usage both in terms of throughput and data size.

Data Federation

A common scenario with data-backed applications is the requirement to be able to access, query, and analyze data from multiple sources. Historically, though, this has required time and work to transform and transfer the data.

MongoDB Atlas Data Federation solves this. It allows you to directly query Atlas databases, Amazon AWS S3 buckets, and HTTP(s) endpoints together in place, without needing to transform or transfer the data. This allows you to maintain the structure of your data.

You can use different combinations of data sources, using what works for you. You can even convert Atlas data from one or more clusters to a different format and output that data to AWS S3 to be consumed by other teams, enabling faster insights with analytics tools.

Data Lake

Atlas Data Lake is a storage service for extracted data, optimized for analytics. It allows you to set up pipelines (not to be confused with aggregation pipelines) that take snapshots of a collection at the cadence specified, from as often as daily. This allows you to have a history of your data, as it was at the time the snapshot was taken.

These snapshots can then be used to perform analytics, queries, and transformations using Data Federation. You can even use a $out stage to output these results to an AWS S3 bucket for use by other teams, tools, or external customers.

Search

MongoDB Atlas Search is a full-text search solution that seamlessly integrates the open source Apache Lucene search engine with Atlas. It is easy to create a search index on your data without needing to create a separate platform or app for searching. The search queries are built using the standard MongoDB Query API using the $search aggregation stage, so you can use the language you already know which makes it quick and easy to get started. It supports all types of data, not just strings, making it powerful right out of the box.

Search queries even support fuzzy search, autocomplete, and synonyms, so if you make a typo, are feeling lazy, or don't quite write the correct term, it will still find something. Life saver!

You can also create the queries using the Atlas UI, so if you prefer a visual way to set up your search and search indexes, you have that option.

Atlas CLI

The Atlas CLI is a command-line interface tool that allows you to interact with your Atlas deployments and Search, using intuitive and short commands in your terminal of choice.

This also opens up the ability to automate deployments, by being able to install and use the CLI in more scenarios.

Atlas App Services

Atlas App Services is a suite of managed cloud services that help you to build apps, integrate other services, and access your data without operation overheads.

Triggers

When building any sort of real-time or event-based applications, one of the key features you may want is a way to respond to events, such as documents being added, deleted, or updated. Or maybe you want to run tasks against a schedule. This is where database triggers come in.

This is helpful if you have data stored in related documents. In our shopping business example, it might be that orders are stored in one document but also inside a customer's document. A trigger could be used to update the customer document, each time there is a new order transaction.

Maybe you want to be notified of a change. For example, if a document is updated in a specific collection, send an email.

Triggers are also good if you want to make calls to an API when something happens, or even when you want version history – for example, changing a last updated value when a document is changed, to help keep record of when changes were made.

Functions

Serverless functions have been around for a while in different cloud providers – for example, Microsoft Azure has Azure Function and AWS has AWS Lambda.

Atlas functions allow you to write and execute server-side logic for your app in JavaScript using ES6+. It is even possible to add external dependencies, allowing you to take advantage of NPM packages to enhance your functionality. Of course, functions also allow you to interact with a MongoDB Atlas database so you can perform CRUD (create, read, update, and delete) actions.

Functions are great for pieces of short-lived (capped at 150 seconds of runtime per request) logic that don't require a full application. It could be that you want to use multiple services, adapt for the current user, or separate some implementation from your application.

You can even set up webhooks so you can call the function externally. The first function I wrote after joining MongoDB was, in fact, a good example of a use case for functions and webhooks.

In my team, we like to share the weather in our location with each other every day, as we are purely remote. So, I wrote a function that calls the OpenWeatherAPI for our locations, maps the weather to a relevant emoji, and then returns this as a JSON object, while also storing it to a collection so we can make a visual dashboard. I then set up a webhook for this that can be called from a Slack application so that we can call a command inside Slack to get the weather for the team! Figure 2-4 illustrates this:

Figure 2-4. *Slack output result from calling Atlas function showing weather for the team and matching emoji*

Another example you see in the sample code from the official Getting Started with Realm tutorial combines triggers and functions. When creating an account from within the app, pressing the "Create Account" button calls a function that takes the entered username and password and uses it to create a new document in the user's collection.

Authentication

When storing data in an application that involves any kind of users, one of the first things you think of is authentication – making sure that the person is who they say they are and ensuring they only see data applicable to them.

Atlas App Services supports role-based users with data access rules to determine read and write permissions. This also means you can associate requests with a user, allowing for auditing.

It also uses authentication providers to support logging in. Out of the box, authentication supports several authentication providers, such as anonymous users, email/password, API keys, and even OAuth 2.0 through Facebook, Google, and Apple id. If you want to implement another authentication provider, you can define your own custom providers, as well.

Once created for the first time, a user has their own application user created, and by using a partition key, such as the user's unique id, they will only see the data associated with that partition key and nobody else's.

Data API

The *Data API* is another great product on offer. Although MongoDB supports drivers in many languages – including C#, NodeJS, Python, and Java – sometimes, you don't always need a driver, or just want to make calls to your database over HTTP.

For example, you might have an IoT sensor that gathers information from the environment and logs this to a JSON file. If you want to store these as documents in a collection in Atlas, creating a whole app with a driver to log this information might be overkill.

This is where the Atlas Data API is so powerful. It provides an HTTP endpoint that can be called to carry out the CRUD operations without needing a driver.

We will discuss more about tools to access and manipulate your data in Chapter 5.

Realm

Realm is an object-oriented mobile database built for making storing, querying, and syncing data much easier, without the need for object-relational mappers (ORMs) or data access objects (DAOs). This is because Realm handles the mapping of the documents to objects in code.

Historically thought of as a mobile database product, Realm has evolved into a database across mobile, web, desktop, and IoT.

Realm started out life as an open source object database management system for iOS and Android. But it could also support cross-platform frameworks such as Xamarin and React Native. In 2019, MongoDB purchased Realm.

Realm uses local databases on the device to store the data, allowing applications to continue working when the device is offline or connectivity is poor. This is also linked to a MongoDB Atlas database.

Because your Realm application is pointing at an Atlas database as well, you can reuse existing data where applicable, saving even more time and allowing data consistency between new and existing projects. Plus, because it is using the document data model, you get the scalability, flexibility, readability, and many other advantages it delivers.

Realm also has SDKs for a variety of languages and frameworks, such as Android, Swift, .NET, Node.js, React Native, Kotlin Multiplatform, Web, and Flutter.

These SDKs give you the ability to interact with, and take advantage of, the products in Atlas App Services, including authentication.

Device Sync

There is also Atlas Device Sync, a feature allowing automatic syncing of the data between a local Realm database and an Atlas cloud database as soon as the devices are online again.

Device Sync handles updates and conflict resolution in the background, so developers are free to concentrate on their own functionality, instead of having to code in network and error handling.

Charts

As a huge Xamarin and .NET fan, Realm is one of my favorite features of MongoDB, but I must also mention my other favorite, Charts. Charts allows you to easily create, share, and embed visualizations from your data in Atlas and Atlas Data Lake, for free, with automatic updating in response to data changes. You don't need to use a third-party graphing library; you can create pretty, collaborating dashboards and charts and use the Charts Embedding SDK or iFrame to add them to your application.

As shown in Figure 2-5, Charts has a UI within the cloud console, just like Atlas, and you can select your data source, chart type, and what fields you want on each axis and have pretty graphs within a few clicks.

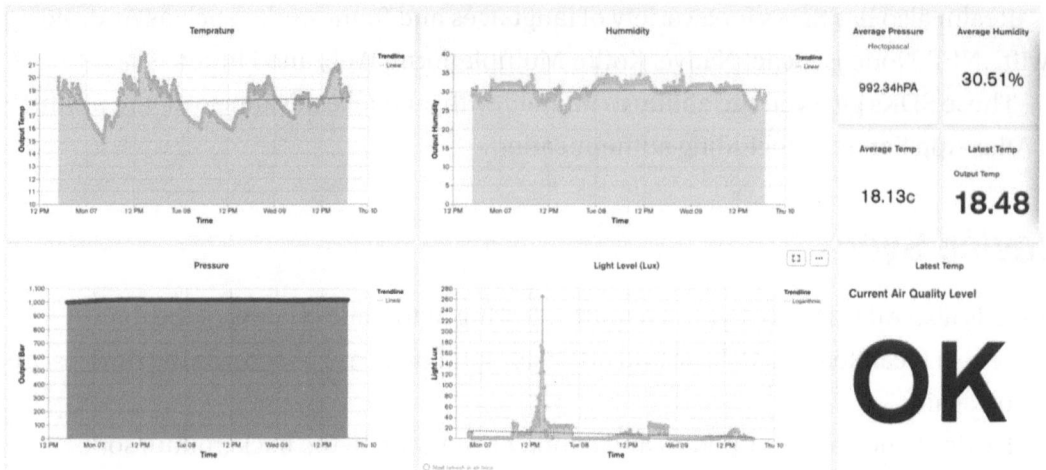

Figure 2-5. *MongoDB Charts dashboard showing graphs for various IoT environment sensors, including temperature and humidity*

A perfect use case for Charts would be the IoT sensor example mentioned earlier. With all this data being logged continually from the sensors, this can quickly stack up to be a lot of documents with information packed in. The ideal solution is graphs to help visualize the data and how it changes.

Summary

- MongoDB has a lot of products, making it a comprehensive application data platform.

- Atlas is not just the core database as a service. It has a variety of tools, including search, triggers, Data Lake, and aggregation.

- Realm is a powerful edge-to-cloud, back-end service providing mobile and web data storage solution, linked to your new or existing Atlas cluster.

- It provides a database with online/offline sync but also functions, user authentication, and static hosting.

- Charts allows for easy creation of visualizations of your data in a variety of charts, with the ability to easily embed into your application with iFrame or the Embedding SDK.

Phew, what a journey! Now that we have learned all about MongoDB and the amazing breadth of products available, it is time to move onto Part 2 and the start of interaction. Join me in Chapter 3 as we create our MongoDB account using your authentication method of choice!

PART II

Setting Up MongoDB

CHAPTER 3

Creating an Account

Well, this is exciting! Now that we have learned more about MongoDB Atlas and the wide variety of features and products on the platform, it is time to go ahead and create our first account so we can start to use it.

In this chapter, we will go through how to create your first MongoDB Account. This could be with your email address and password, Google Account, or even GitHub.

Creating Your MongoDB Account

N.B. This flow and screenshots are accurate at the time of writing. Your experience may vary.

1. Visit `https://account.mongodb.com/account/login` to access the MongoDB Atlas cloud console. Figure 3-1 shows what you will see when you visit the MongoDB Cloud Landing Page.

2. Click "Don't have an account? Sign up" on the left side of the page.

© Luce Carter 2024
L. Carter, *Beginning MongoDB Atlas with .NET*, https://doi.org/10.1007/978-1-4842-9550-2_3

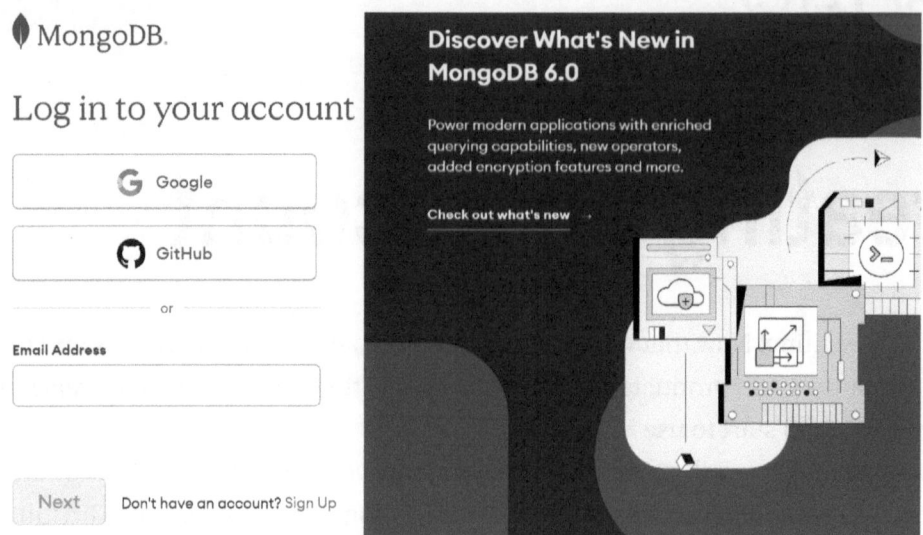

Figure 3-1. *MongoDB Cloud Landing Page*

You now have three choices for how you wish to create your account. You can create an account using a Google Account, such as a Gmail account, a GitHub account, or one hosted using Google Workspace, or enter your details and create a new account using your email address. Let's have a look at both methods.

Google Account

The following steps will help you sign up if you choose to use Google as your authenticator:

1. On the "Create your account" page, as shown in Figure 3-2, select the "Google" button.

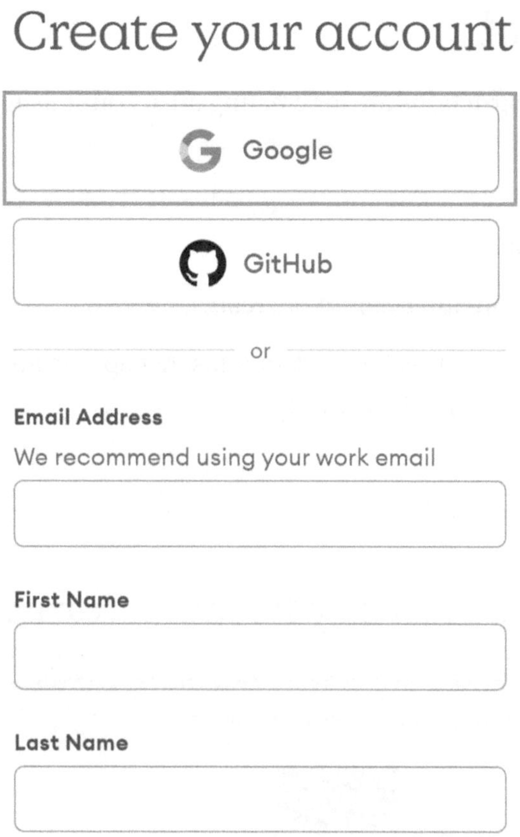

Create your account

G   Google

◯   GitHub

or

Email Address
We recommend using your work email

First Name

Last Name

Figure 3-2. *Create your account page with Google button visible*

2. Select your chosen Google Account or create one in the box that appears.

3. You will be redirected back to Atlas once you are logged in. Check the box to accept the Privacy Policy and Terms of Service and click "Submit."

GitHub Account

The following sets of instructions are for creating an account using an existing GitHub profile:

1. On the "Create your account" page, select GitHub as your sign-up option.

2. Login to your GitHub account or create a new one.

3. You will then be redirected back to Atlas. Accept the terms and conditions, as seen in Figure 3-3.

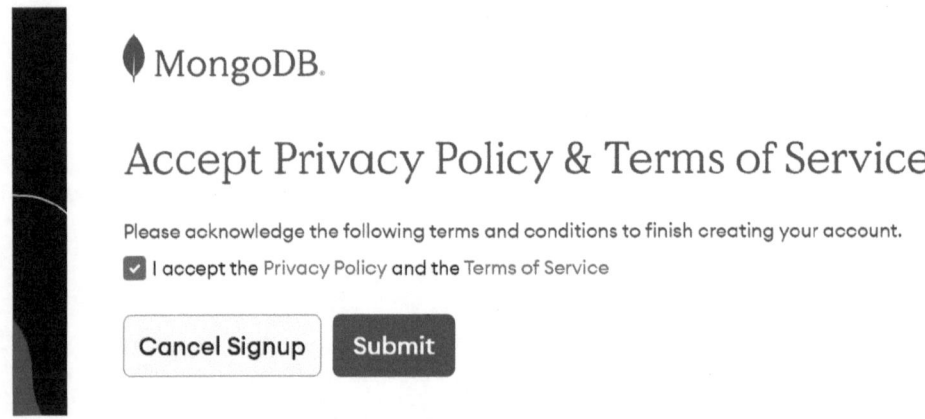

Figure 3-3. *Accept Terms of Service*

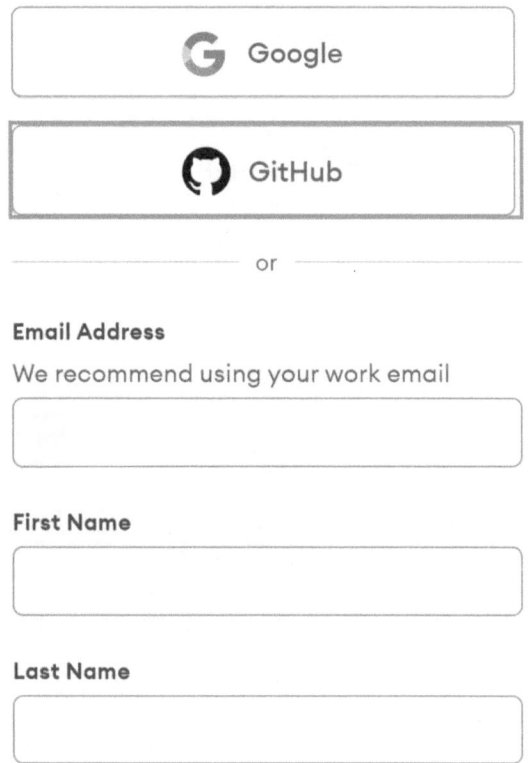

Figure 3-4. *Create your account page with GitHub button visible*

Create a New Account

The following sets of instructions are for creating a new account on Atlas, without using Google:

1. Fill out your details in the boxes provided, as shown in Figure 3-5, check the box to accept the Privacy Policy and Terms of Service, and click "Sign up."

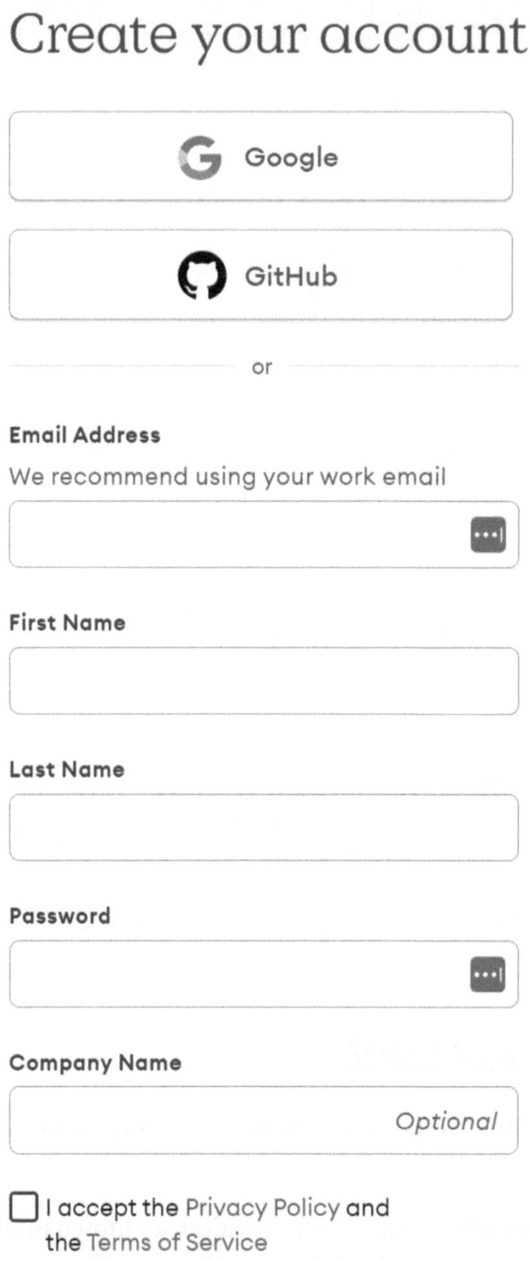

Create your account

G   Google

GitHub

— or —

Email Address

We recommend using your work email

[   ] [···]

First Name

[   ]

Last Name

[   ]

Password

[   ] [···]

Company Name

[*Optional*]

☐ I accept the Privacy Policy and
the Terms of Service

Figure 3-5. *Enter your details to create an account in Atlas*

2. You will then see a screen requesting that you verify your email, as shown in Figure 3-6. Check your inbox for an email from MongoDB Cloud and click the "Verify Email" button within the email. Be aware it may be delivered to your junk/spam folder.

Great, now verify your email

Check your inbox at ██████████████████████████████ and click the verification link inside to complete your registration. This link will expire shortly, so verify soon!

Don't see an email? Check your spam folder.

Link expired? Resend verification email

Figure 3-6. *Verify your email address*

3. As seen in Figure 3-7 below, the link will take you to a page to confirm that your email address has been verified. Click "Continue."

Email successfully verified!

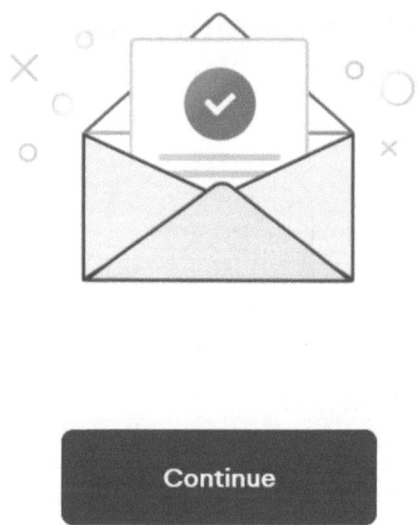

Continue

Figure 3-7. *Email successfully verified*

Finalizing Your New Account

Once you reach this point, the steps are the same regardless of how you created your account.

1. You will be met with a Welcome to Atlas screen, as seen in Figure 3-8. Answer the questions about yourself and your project. What you put here doesn't matter too much. The main thing is to not forget that our preferred language is C#/.NET. Then click "Finish."

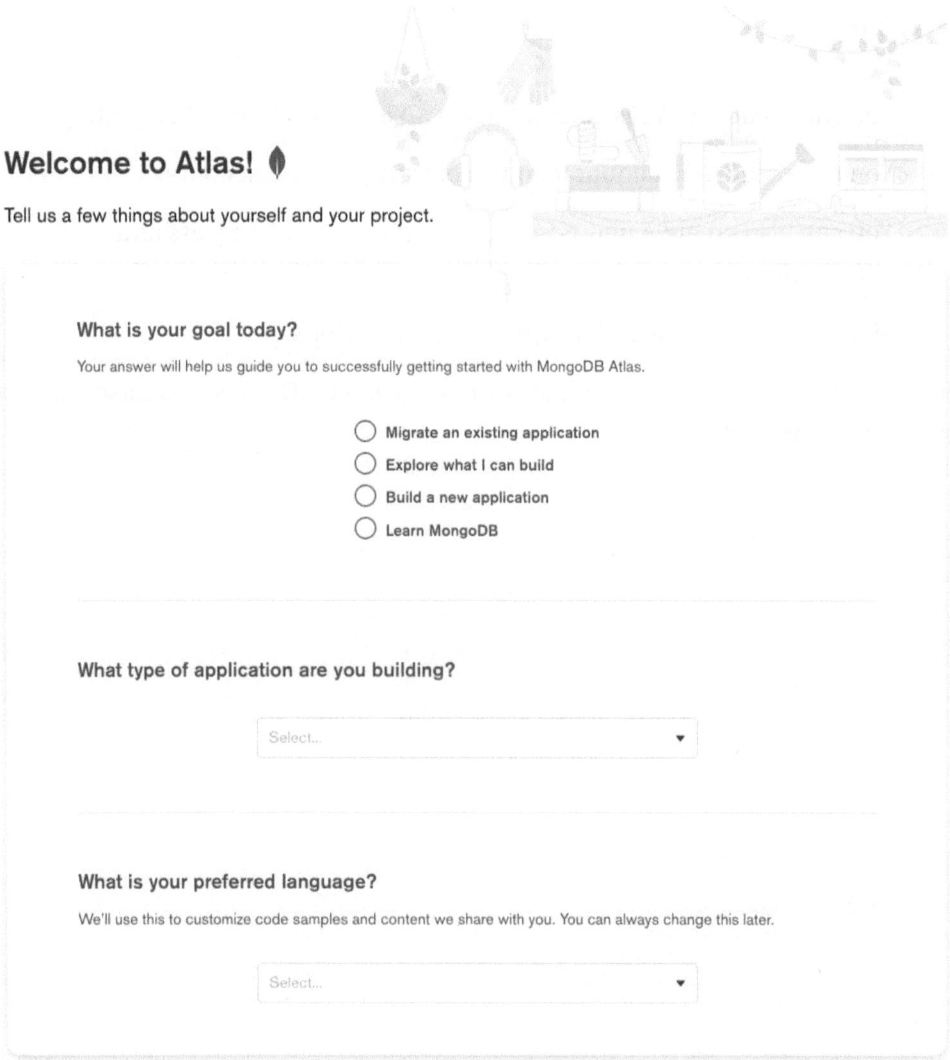

Welcome to Atlas! ♦

Tell us a few things about yourself and your project.

What is your goal today?

Your answer will help us guide you to successfully getting started with MongoDB Atlas.

○ Migrate an existing application
○ Explore what I can build
○ Build a new application
○ Learn MongoDB

What type of application are you building?

Select...   ▼

What is your preferred language?

We'll use this to customize code samples and content we share with you. You can always change this later.

Select...   ▼

Figure 3-8. *Questions about you and your project*

From here, you will reach a screen about deploying your database. We will cover this in Chapter 4, so keep reading as we deploy our first ever free database cluster in MongoDB Atlas.

Summary

- You can create an Atlas account using an existing Google Account.

- You can create an account using an existing GitHub account.

- You can also create a new account using your email address and contact details.

- Signing up is simple and only requires a few steps.

On to Chapter 4, where we will create our first MongoDB Atlas Database, all completely free, forever!

CHAPTER 4

Creating Your First Cluster and Loading Sample Dataset

We're getting closer to writing code! The goal of this chapter is to create your first database cluster and load the sample dataset. The sample dataset is amazing. It is around 350MB of totally free data, available to you out of the box, that you can load with the click of a button.

Once loaded, a number of databases and collections will be created that include numerous documents of varying shapes and sizes for you to explore.

Creating Your First Cluster

In this first section, we will go ahead and create our first database cluster in MongoDB Atlas.

1. Visit `https://cloud.mongodb.com` to access the MongoDB Atlas cloud console. If you're not logged in, log in using the account you created in Chapter 3.

2. As shown in Figure 4-1, the first time you load a project within Atlas, it won't have any databases, so you will be met with a screen with a big button that says, "Build a Database." Click it.

51

L. Carter, *Beginning MongoDB Atlas with .NET*, https://doi.org/10.1007/978-1-4842-9550-2_4

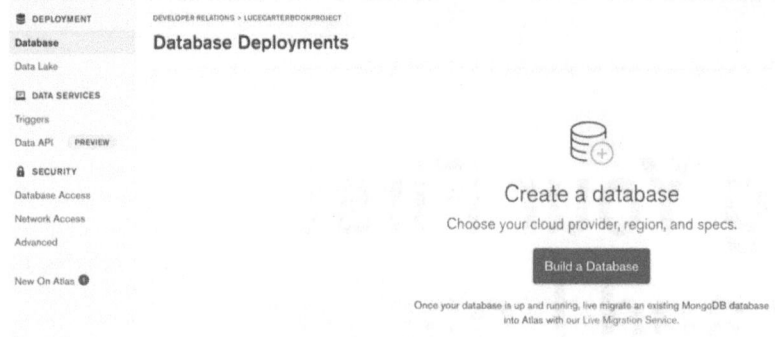

Figure 4-1. *New project in Atlas. Click the button to build your first database*

3. You will then be met with a choice of what cluster type to pick: ***Serverless***, ***Dedicated***, or ***Shared***. We want Shared, which is the free forever tier. To create your first database, click the "Create" button under the ***Shared*** heading (Figure 4-2).

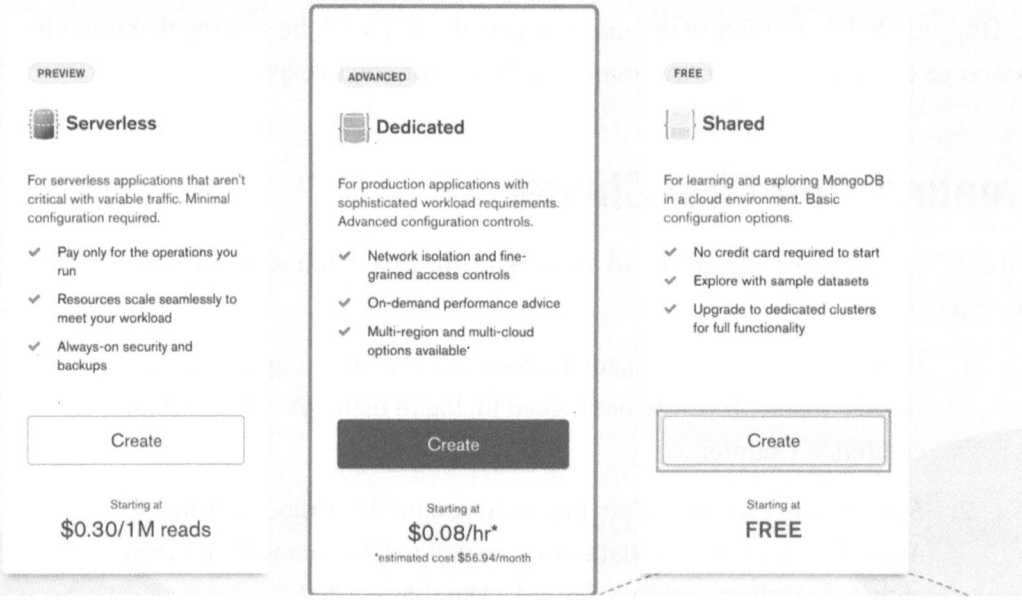

Figure 4-2. *Selecting a Shared cluster*

4. Next is your chance to pick a cloud provider and region of choice. In Figure 4-3 you can see AWS is selected. Once the Cloud Provider has been selected, choose a Cloud Provider region.

    MongoDB recommends picking the region that is geographically nearest to either you or where your application will be hosted.

Figure 4-3. *Selecting Cloud Provider & Region*

5. Below the cloud provider section is an area where you can make some configuration changes to your cluster. For now, we are going to use the default values as this will match the code you will see later on. So, go ahead and click "Create Cluster" below this (Figure 4-4).

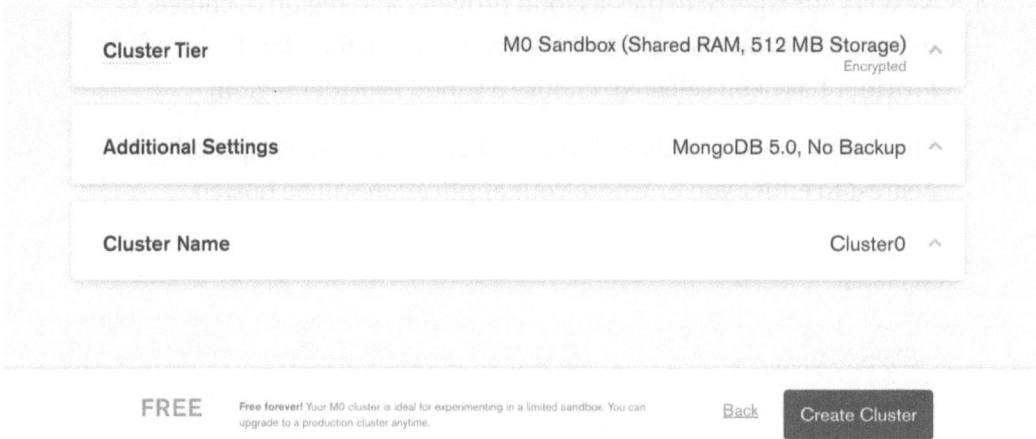

Figure 4-4. *Create Cluster with default settings*

6. In the Security Quickstart window, enter a secure username and
    password that will be used going forward to access the database,
    and click "Create User" (Figure 4-5).

Security Quickstart

To access data stored in Atlas, you'll need to create users and set up network security controls. Learn more about security setup

1 How would you like to authenticate your connection?

Your first user will have permission to read and write any data in your project.

| Username and Password | Certificate |

Create a database user using a username and password. Users will be given the *read and write to any database*
privilege by default. You can update these permissions and/or create additional users later. Ensure these credentials
are different to your MongoDB Cloud username and password.

Username

Enter username

Password

Enter password   Autogenerate Secure Password   Copy

Create User

Figure 4-5. *Set up Username and Password in Security Quickstart*

7. The next stage in the Security Quickstart is to add your IP address to the list of allowed locations by clicking "Add My Current IP Address." Then click "Finish and Close" (Figure 4-6). You need to configure this because MongoDB adds security by default around data access.

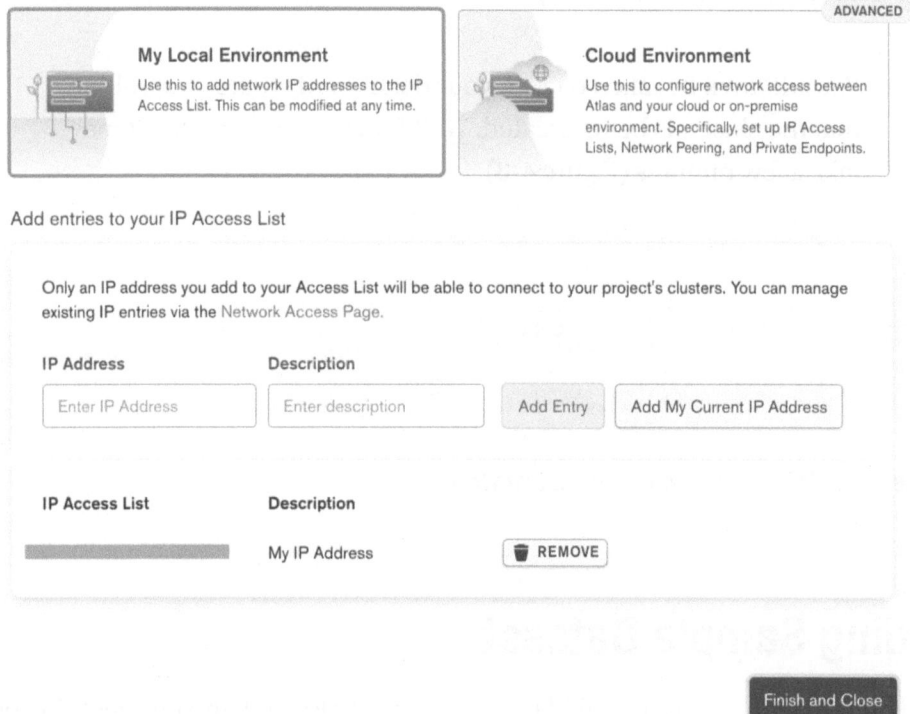

Figure 4-6. *Add IP Address to Access List*

8. We are now done configuring your first ever database in Atlas, so click "Go to Databases" (Figure 4-7).

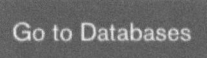

Congratulations on setting up access rules!

You will now be able to connect to your deployments. You can continue to add and update access rules in Database Access and Network Access.

☑ Hide Quickstart guide in the navigation. You can visit Project Settings to access it in the future.

Go to Databases

Figure 4-7. *Configuration is complete*

9. You will then be taken back to your dashboard where you will see that the cluster is being created. Wait for this to complete – it may take a few minutes (Figure 4-8).

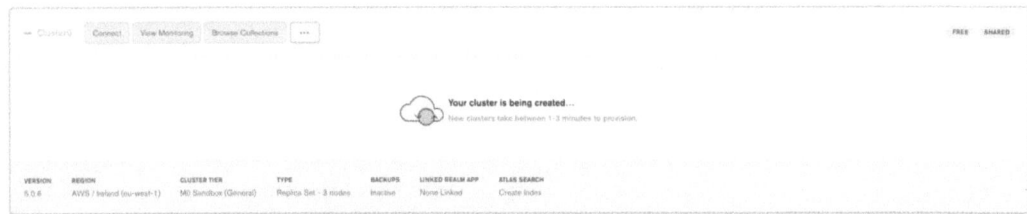

Figure 4-8. *Wait for cluster to be created*

Loading Sample Dataset

We have access to a very thorough dataset that can be loaded into our new cluster from MongoDB Atlas' user interface.

1. From the database deployment dashboard, you will see your new cluster will now be available to you. Click the ellipses button, or "...", to access some further options for your cluster (Figure 4-9).

Figure 4-9. *Click ellipses button from Cluster dashboard*

2. Click "Load Sample Dataset" from the dropdown menu (Figure 4-10).

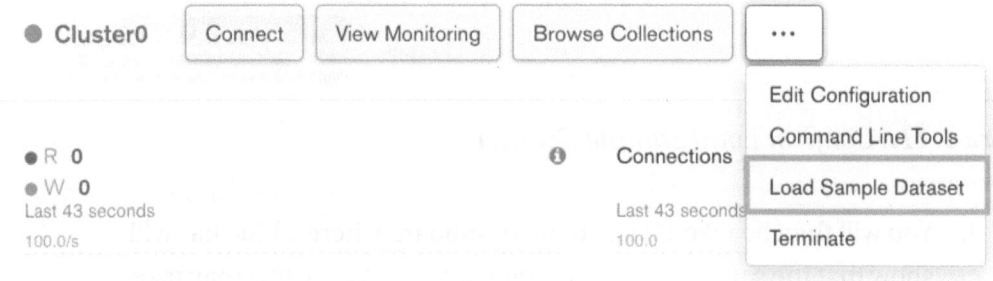

Figure 4-10. *Select Load Sample Dataset*

3. Confirm that you want to load the sample dataset by clicking "Load Sample Dataset" (Figure 4-11).

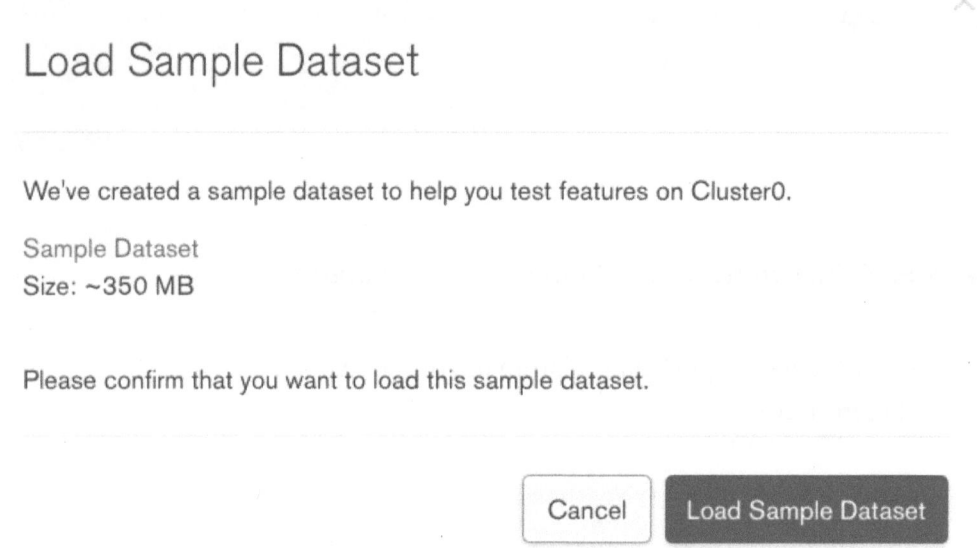

Figure 4-11. *Confirm Load Sample Dataset*

4. You will then be taken back to the dashboard where a blue bar will show that the sample dataset is being loaded. Again, this may take a few minutes (Figure 4-12).

Figure 4-12. *Wait for sample dataset to be loaded*

Summary

Things are really coming together. You now have a brand-new cluster with sample data available to play around with.

- Creating a cluster is as simple as clicking a button in Atlas and following a short wizard.

- You can create different types of clusters, but one of those is an M0 cluster which is free forever and great for getting started.

- Security like authentication and IP access restriction lists are included by default.

- Every cluster has access to a sample dataset that can be loaded for free.

In Chapter 5, we will look at different ways you can browse your data to not only see what is in the sample dataset we just added but how you can also manipulate this data with CRUD operations and aggregations.

CHAPTER 5

Browsing Your Data

MongoDB Atlas offers a wide variety of tooling to help browse and interact with your cloud-hosted data. In this chapter, we will look at what options are available, including examples of basic read and update functions. By the end, you will understand what tools are available and be able to make an informed choice as to which one works best for you.

Atlas UI

The Atlas UI, also referred to as the cloud console, is often the most common way users access and interact with their Atlas services and clusters. It's accessed from the browser, so no third-party tools are required. From the Atlas UI, you can create databases, manage existing databases and collections, browse your data, and carry out a variety of functions from adding, deleting, and updating documents to performance advice to search index management to security.

Reading Data

While in the Atlas UI and on the dashboard for your database cluster, you can click Browse Collection, as shown in Figure 5-1, to start accessing your data.

© Luce Carter 2024
L. Carter, *Beginning MongoDB Atlas with .NET*, https://doi.org/10.1007/978-1-4842-9550-2_5

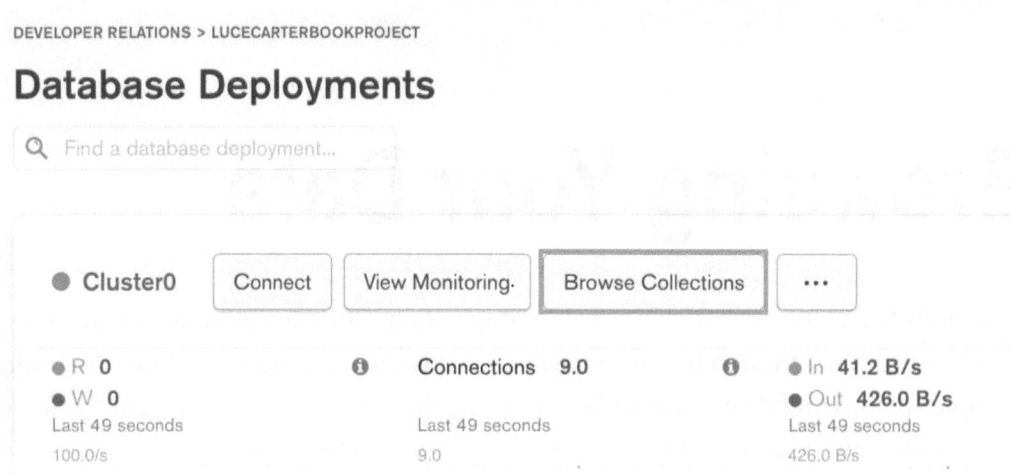

Figure 5-1. *Atlas UI Browse Collections button*

You will then be taken to your database, specifically the collections view, where you can see a list of the collections you have in your database. Figure 5-2 shows a database with the sample data collections we added in Chapter 4.

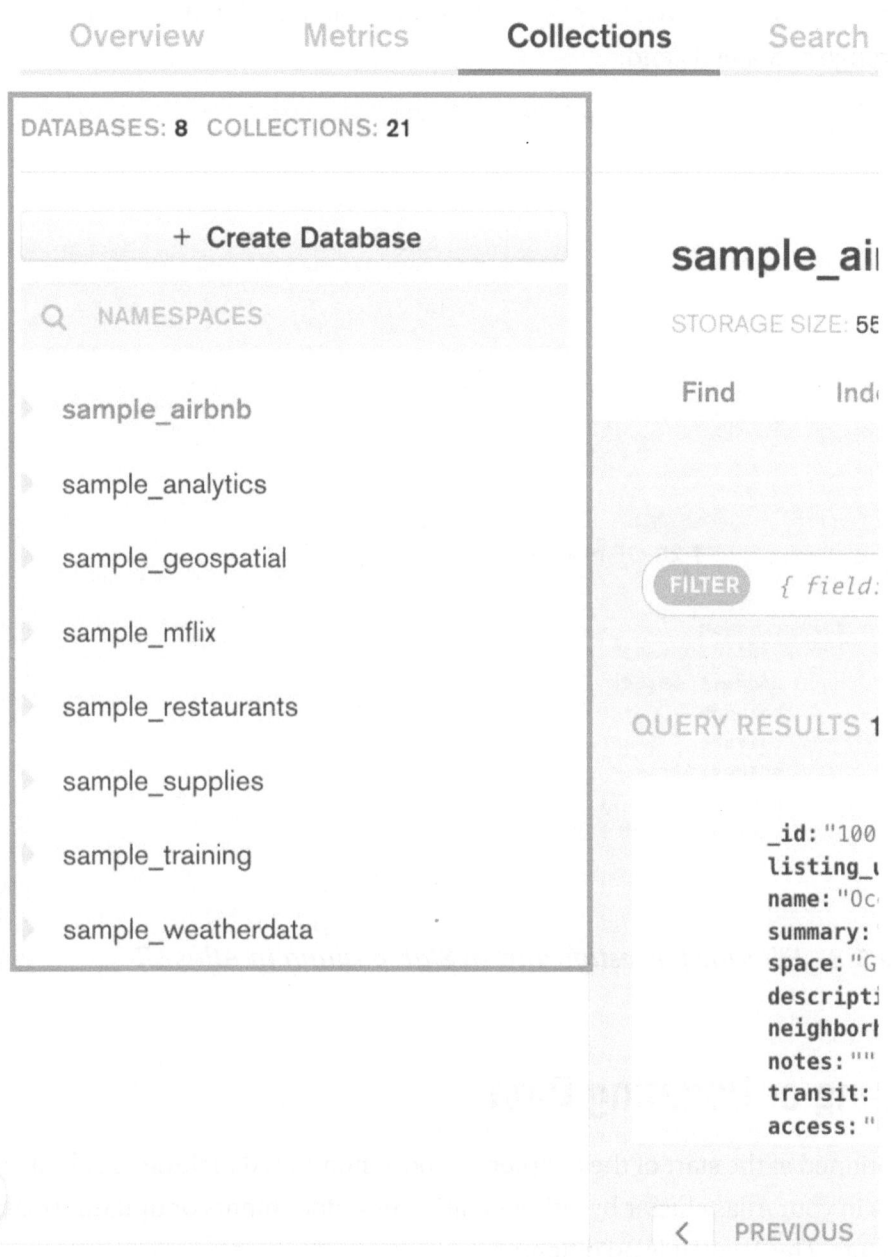

Figure 5-2. *Atlas UI showing collections from sample dataset*

The collections view is a great way to see your data, including from the sample dataset, as you can even run queries against the data in the selected collection from inside the browser.

Figure 5-3 shows the filter box inside the same UI view in Atlas, with an example of a find query, searching in the sample_restaurants database for all restaurants located in the borough of Staten Island.

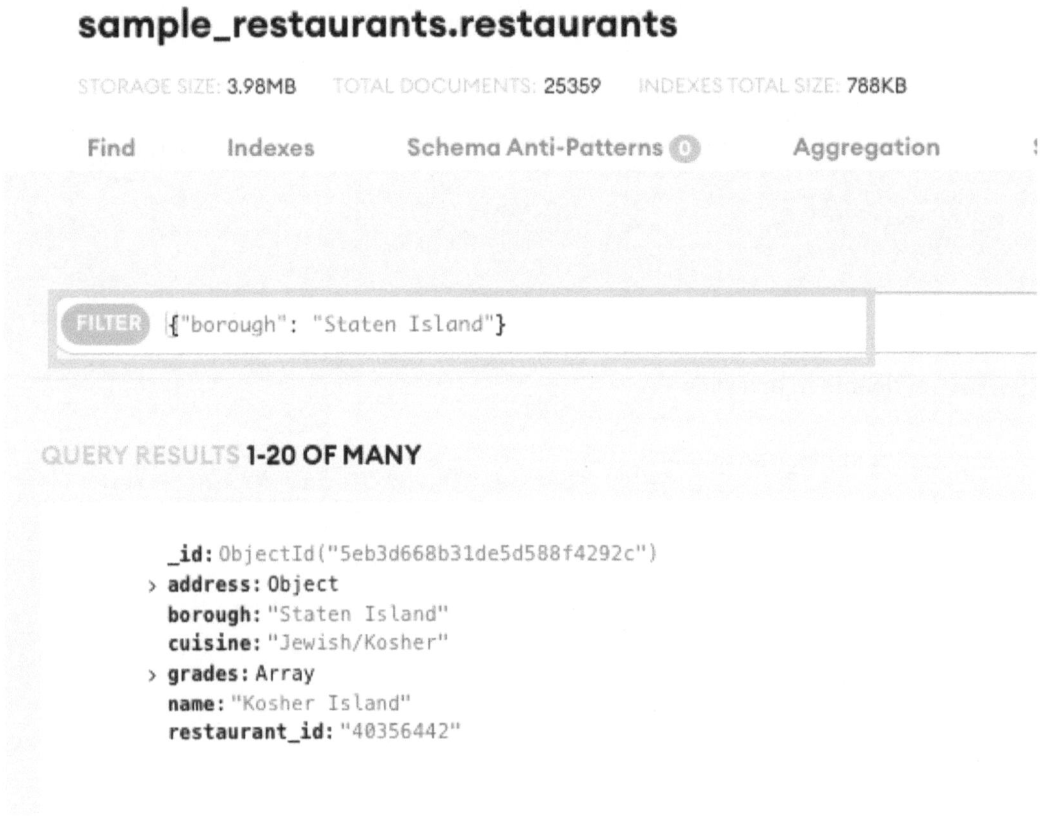

sample_restaurants.restaurants

STORAGE SIZE: **3.98MB**   TOTAL DOCUMENTS: **25359**   INDEXES TOTAL SIZE: **788KB**

Find   Indexes   Schema Anti-Patterns ⓪   Aggregation   ⁝

FILTER   {"borough": "Staten Island"}

QUERY RESULTS **1-20 OF MANY**

```
    _id: ObjectId("5eb3d668b31de5d588f4292c")
  > address: Object
    borough: "Staten Island"
    cuisine: "Jewish/Kosher"
  > grades: Array
    name: "Kosher Island"
    restaurant_id: "40356442"
```

Figure 5-3. *Filtering for restaurants in Staten Island in Atlas UI*

Creating or Updating Data

As mentioned at the start of the chapter, all tools mentioned include the ability to modify the data in your Atlas cluster by either creating new documents or updating existing documents. The Atlas UI is no different.

Figure 5-4 shows the location of the insert document button in the UI, allowing you to create new documents in a collection without ever leaving the browser.

Figure 5-4. *Insert Document button inside Atlas UI*

This will open a modal box, allowing you to insert a document. The default view, as seen in Figure 5-5, shows the UI where you can select the data type on the right, enter fields and values, and add new fields.

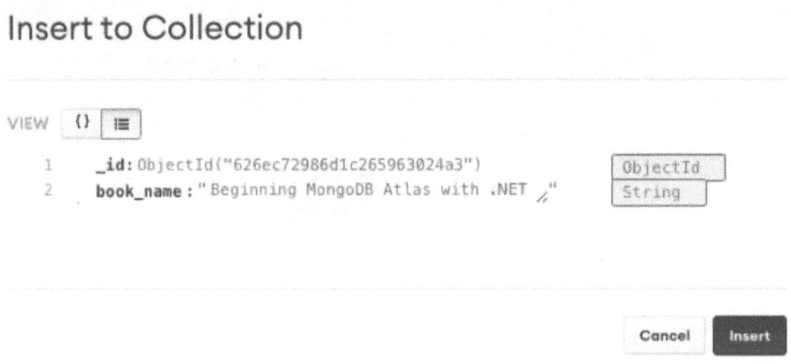

Insert to Collection

Figure 5-5. Insert a document to a collection in Atlas UI

If you prefer working with JSON documents however, it's possible to change the view and use traditional JSON to insert a document, as seen in Figure 5-6.

What is clever as well is if you add a field in one view and then switch views, it will persist that new field between views, even if you haven't clicked Insert yet to add the new document and persist the values.

Figure 5-6. Insert a document to a collection in Atlas UI using JSON

Regardless of which view you choose to add fields to a new document for your collection, clicking the green Insert button in the bottom right of the modal window will instantly add the new document to your collection.

It's also possible to update existing documents from within the Atlas UI. On each document, when you hover your mouse cursor over any part of the document area in the UI, a toolbar will appear in the top right of that area with options to Edit, Copy, Duplicate, or Delete the document, as seen in Figure 5-7.

Figure 5-7. *Edit button on document in Atlas UI*

Clicking Edit will load the document in the default UI view, as seen in Figure 5-8. When updating an existing document, it's currently not possible to edit it in the JSON view.

Figure 5-8. *Updating an existing document*

Deleting Data

Deleting data in the Atlas UI is as simple as clicking the trash bin icon that appears in the top-right menu, as shown in Figure 5-9. Clicking this will display a banner at the bottom warning of the deletion, and you just click Delete again to instantly remove the document from the collection.

Figure 5-9. *Delete a document button in the Atlas UI*

Though you can manage your data directly through the Atlas UI, some people prefer to use command-line interface (CLI) tools instead. This is where the next tool comes in, Mongosh.

Mongosh

MongoDB's CLI shell tool is called "Mongosh." It's a fully functioning Node.js REPL and allows you to interact with your MongoDB database and collections from the command line.

You can perform all the operations you would expect, including CRUD operations and aggregation pipelines. You can even write scripts using JavaScript that manipulate data in your database or perform administrative tasks.

Once Mongosh is installed, to begin using it with your Atlas cluster, you'll first need to establish a connection using a connection string. This can be done via the Atlas UI by first selecting Connect, as shown in Figure 5-10, then selecting your connection method of choice.

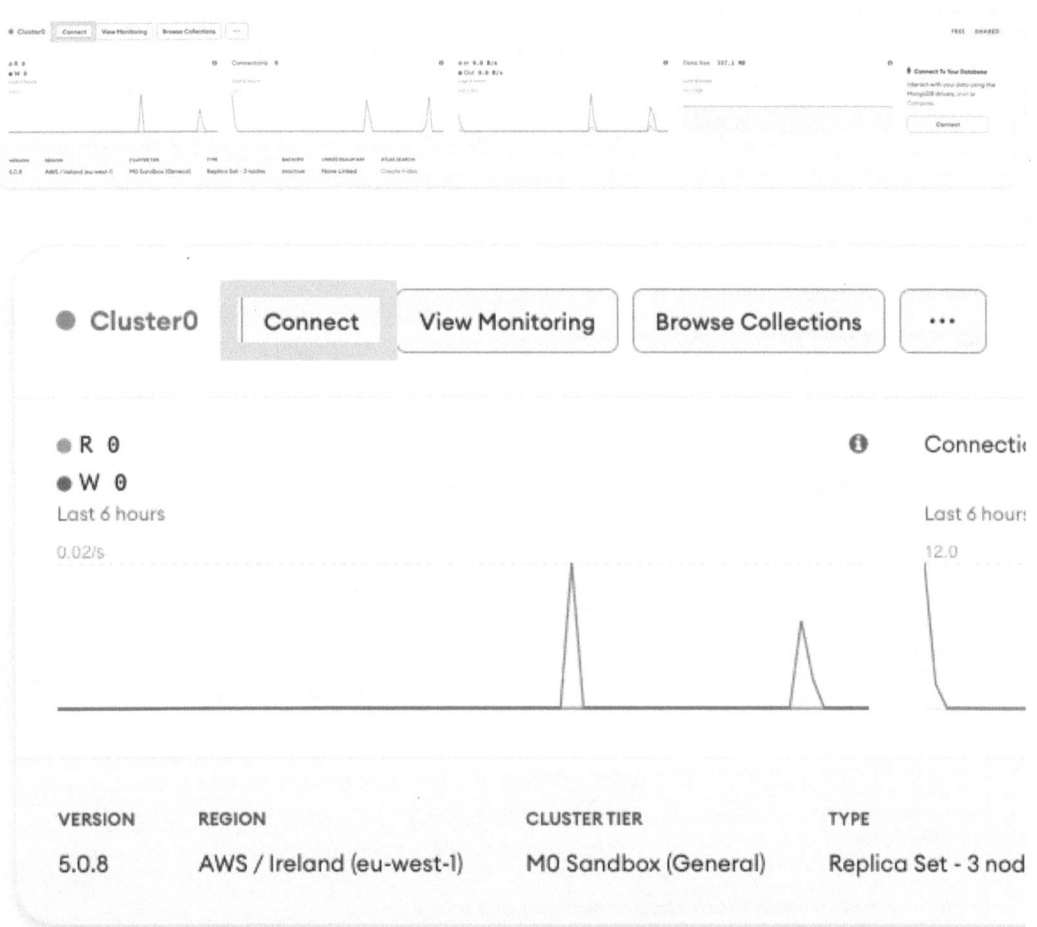

Figure 5-10. *The connect button for a cluster in Atlas UI*

If you select Mongosh, it will give you two options to select: either that you do not have the MongoDB Shell installed or that you do. Either tab will include instructions on how to connect to your selected cluster from Mongosh. It will give you a unique connection string, including the mongosh command, so you can copy and paste it directly into your CLI, as seen in Figure 5-11.

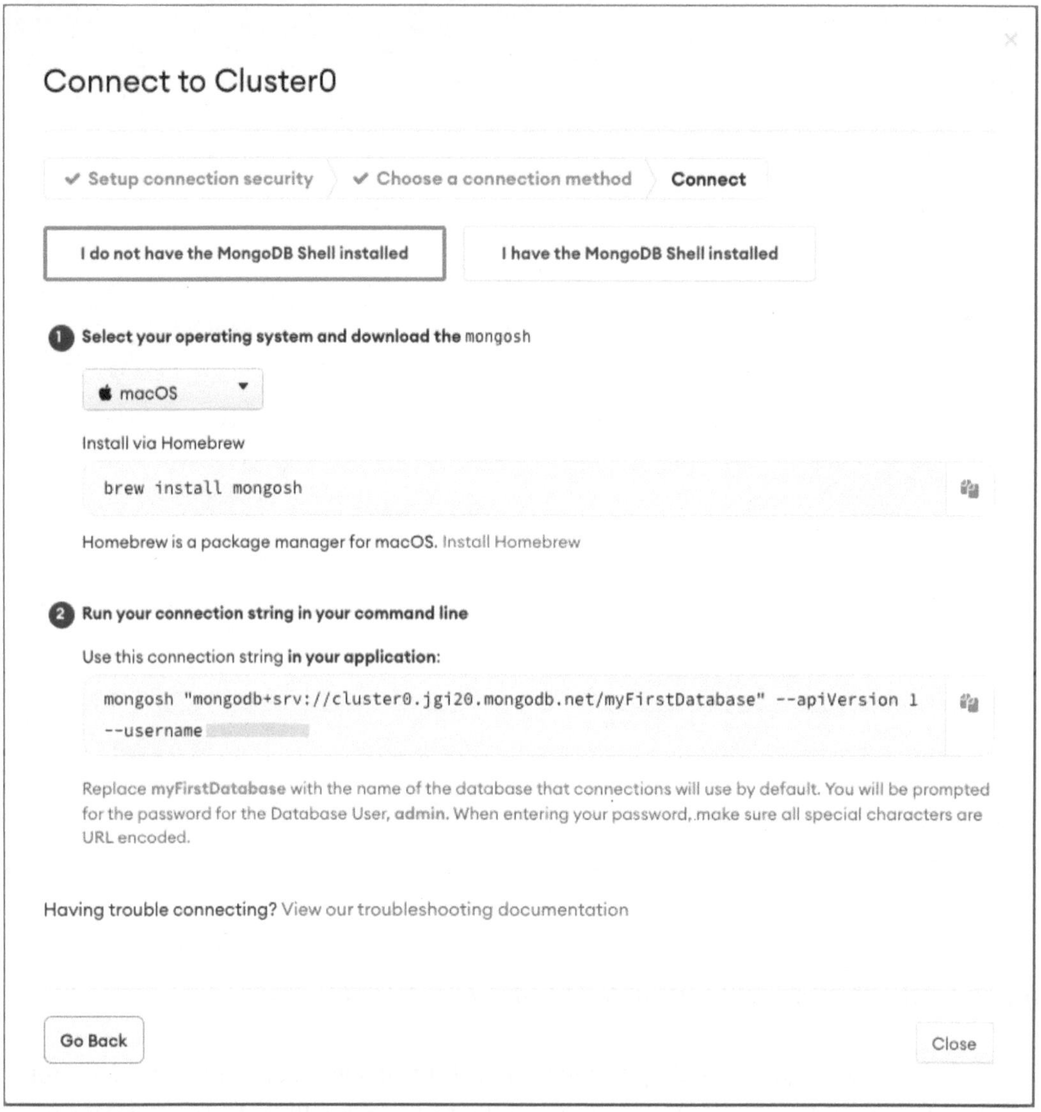

Connect to Cluster0

✔ Setup connection security ✔ Choose a connection method **Connect**

| I do not have the MongoDB Shell installed | I have the MongoDB Shell installed |

1 Select your operating system and download the mongosh

🍎 macOS ▾

Install via Homebrew

```
brew install mongosh
```

Homebrew is a package manager for macOS. Install Homebrew

2 Run your connection string in your command line

Use this connection string **in your application:**

```
mongosh "mongodb+srv://cluster0.jgi20.mongodb.net/myFirstDatabase" --apiVersion 1
--username
```

Replace **myFirstDatabase** with the name of the database that connections will use by default. You will be prompted for the password for the Database User, **admin.** When entering your password, make sure all special characters are URL encoded.

Having trouble connecting? View our troubleshooting documentation

Go Back   Close

Figure 5-11. *Connection instructions for MongoDB Shell, a.k.a. Mongosh, in the Atlas UI*

Once you successfully log in with your connection string, you will be met with a prompt like Figure 5-12 which is where you can start to interact with your cluster and take advantage of the powers of Mongosh, the MongoDB Shell.

```
Current Mongosh Log ID: 626ee85faed077f19fb12e4a
Connecting to:        mongodb+srv://cluster0.jgi20.mongodb.net/myFirstDatabase?appName=mongosh+1.3.1
Using MongoDB:        5.0.8 (API Version 1)
Using Mongosh:        1.3.1

For mongosh info see: https://docs.mongodb.com/mongodb-shell/


To help improve our products, anonymous usage data is collected and sent to MongoDB periodically (https://www.mongodb.com/legal/privacy-policy).
You can opt-out by running the disableTelemetry() command.

Atlas atlas-z2cgvd-shard-0 [primary] myFirstDatabase>
```

***Figure 5-12.** Initial prompt after successful connection to cluster in Mongosh*

Reading Data

Reading data is very simple, as you can take advantage of the powerful MongoDB Query API discussed in earlier chapters. Figure 5-13 shows listing all databases, selecting to use that database, and then doing a simple find query on a collection in that database. This find query syntax will look familiar from the Atlas UI. Note that it uses the find method. This will return any documents that match those criteria. If you want to find only the first one that matches your search criteria, use the findOne method instead.

```
Atlas atlas-z2cgvd-shard-0 [primary] myFirstDatabase> show dbs
book_sampledata         41 kB
sample_airbnb         54.7 MB
sample_analytics      9.56 MB
sample_geospatial     1.48 MB
sample_mflix          56.3 MB
sample_restaurants    6.94 MB
sample_supplies       1.18 MB
sample_training       55.3 MB
sample_weatherdata     2.9 MB
admin                  344 kB
local                 2.11 GB
Atlas atlas-z2cgvd-shard-0 [primary] myFirstDatabase> use book_sampledata
switched to db book_sampledata
Atlas atlas-z2cgvd-shard-0 [primary] book_sampledata> db.books.find({"author": "Luce Carter"});
[
  {
    _id: ObjectId("626eddb086d1c265963024a5"),
    book_name: 'Beginning MongoDB Atlas with .NET',
    author: 'Luce Carter',
    publisher: 'Apress'
  }
]
```

***Figure 5-13.** Interacting with Mongosh to find a document in the collection*

Creating Data

Creating data is also simple. The `insertOne` method accepts a JSON-like document structure as a parameter, which represents the document you want to insert. If the insertion is successful, you will see a response like that shown in Figure 5-14.

```
Atlas atlas-z2cgvd-shard-0 [primary] book_sampledata> db.books.insertOne({ "name": "Xamarin.Forms Essentials", "author": "Gerald Versluis", "publisher": "Apress"})
{
  acknowledged: true,
  insertedId: ObjectId("626eeda18049282b16d6f4ef")
}
Atlas atlas-z2cgvd-shard-0 [primary] book_sampledata>
```

Figure 5-14. *Inserting a document using Mongosh*

If you want to insert multiple documents, use the `insertMany` method instead, but pass an array of documents instead of a single document as the parameter.

Updating Data

Updating data from inside Mongosh is also very easy with the MongoDB Query API. You can simply call the `updateOne` method, which takes two arguments. The first is a query to find the document you want to update, and the second is an operator of what change you want to make. In our simple example, as seen in Figure 5-15, we use `$set`, which will set the "name" field to the new value.

```
Atlas atlas-z2cgvd-shard-0 [primary] book_sampledata> db.books.updateOne({"author": "Gerald Versluis"}, {"$set": {"name": "Xamarin.Forms: Essentials"}})
{
  acknowledged: true,
  insertedId: null,
  matchedCount: 1,
  modifiedCount: 0,
  upsertedCount: 0
}
Atlas atlas-z2cgvd-shard-0 [primary] book_sampledata>
```

Figure 5-15. *Calling updateOne inside Mongosh*

It's outside the scope of this book, but you can use all the $ operators from the aggregation pipeline when updating one or many documents in one command.

Deleting Data

Deleting data is equally as simple. The MongoDB Query API, as you have seen, is a very intuitive. If we want to delete one of the documents in the collection, we can call deleteOne, passing in a filter to locate which document to delete. This is often the _id field, but of course you can pass in something more complex.

In Figure 5-16, we delete the first document that has the author "Gerald Versluis."

```
Atlas atlas-z2cgvd-shard-0 [primary] book_sampledata> db.books.deleteOne({"author": "Gerald Versluis"})
{ acknowledged: true, deletedCount: 1 }
```

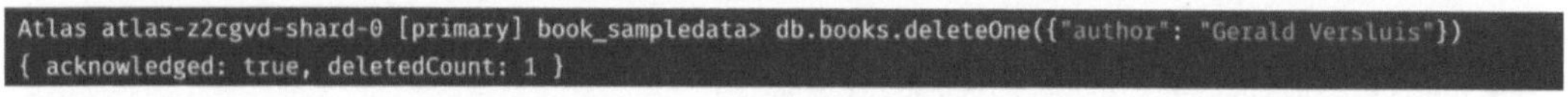

Figure 5-16. *MongoDB Shell deleteOne command*

If you prefer to use a more graphical approach, then the next tool will be perfect: MongoDB Compass.

MongoDB Compass

Compass is MongoDB's free graphical user interface (GUI) application, available on Linux, Mac, and Windows. As well as allowing you to browse your data and make changes to it like in Atlas, with a similar UI, you can also use the embedded aggregation builder against your collections, to help you build the pipeline and visualize how each stage will change the data.

In Figure 5-10, you saw the Connect button in the Atlas UI, which allows you to get the connection details for your chosen method of choice.

We can follow the same steps as in the shell, but instead select Compass as our connection choice, as shown in Figure 5-17.

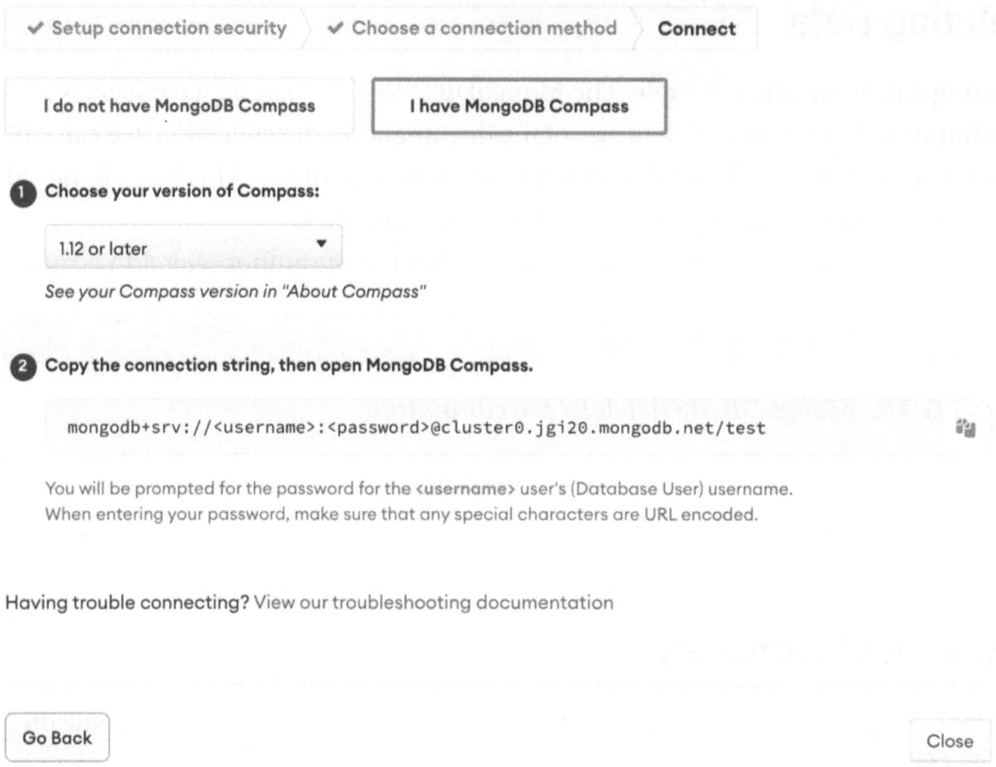

✔ Setup connection security ⟩ ✔ Choose a connection method ⟩ **Connect**

| I do not have MongoDB Compass | | **I have MongoDB Compass** |

1 **Choose your version of Compass:**

1.12 or later ▼

See your Compass version in "About Compass"

2 **Copy the connection string, then open MongoDB Compass.**

mongodb+srv://<username>:<password>@cluster0.jgi20.mongodb.net/test

You will be prompted for the password for the **<username>** user's (Database User) username.
When entering your password, make sure that any special characters are URL encoded.

Having trouble connecting? View our troubleshooting documentation

Go Back   Close

Figure 5-17. *Getting the connection string for Compass inside the Atlas UI*

Reading Data

Reading data within Compass will look very familiar to the Atlas UI. After copying the connection string from your Compass Connection window in the Atlas UI, as shown in Figure 5-17, you can open Compass.

Once you open Compass for the first time, you will be met with a New Connection section, as shown in Figure 5-18.

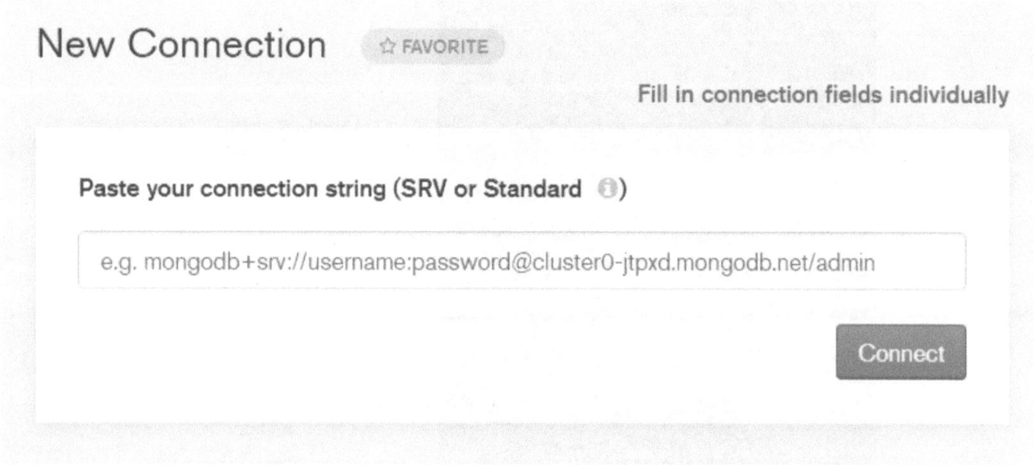

New Connection   ☆ FAVORITE

Fill in connection fields individually

Paste your connection string (SRV or Standard ⓘ)

e.g. mongodb+srv://username:password@cluster0-jtpxd.mongodb.net/admin

Connect

Figure 5-18. *Adding new connection inside Compass GUI tool*

You can then paste your connection string into the New Connection dialog box, updating the username and password values with your own, before clicking connect.

If your details are correct, it will connect and you will be able to see the list of databases on the left, as well as being able to expand them to show collections and browse the documents in that collection, as seen in Figure 5-19.

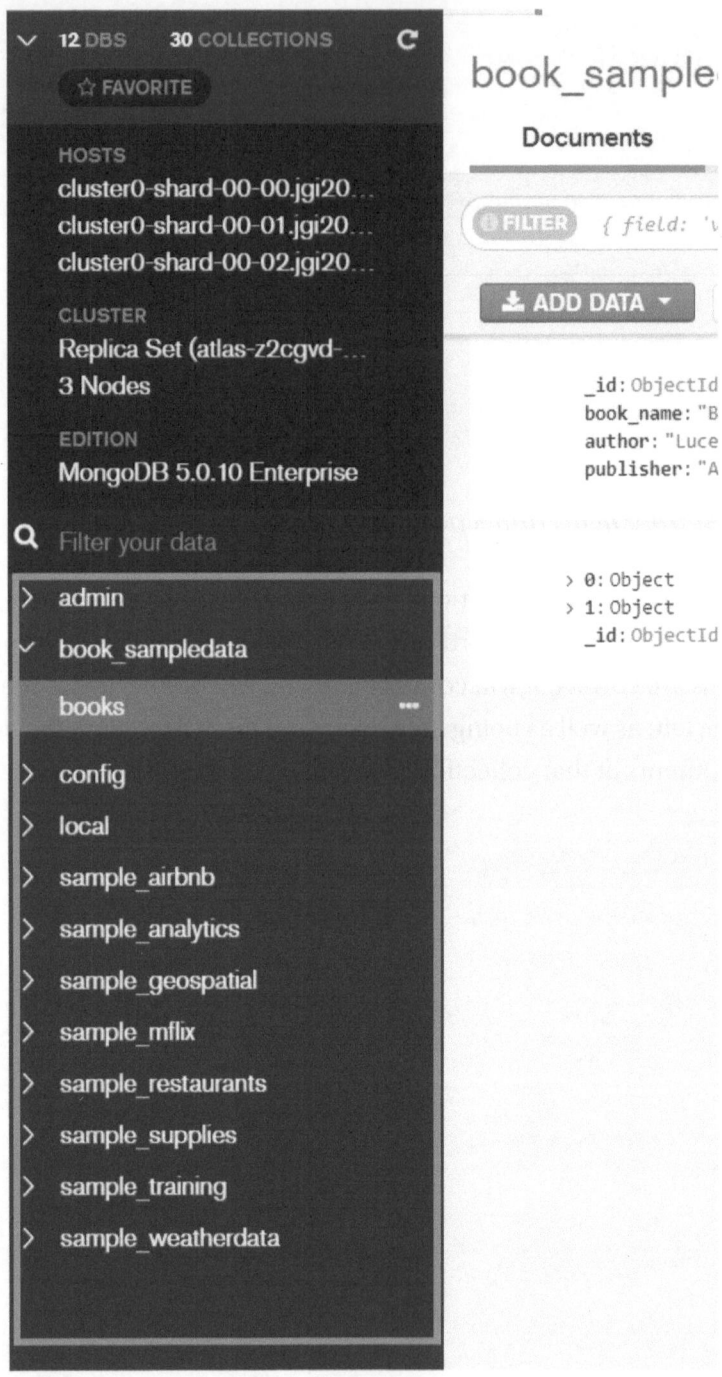

Figure 5-19. *Compass connected to a cluster in Atlas, showing existing databases and collections*

Creating, Updating, or Deleting Data

Creating and manipulating data is also very simple in Compass and almost identical to the Atlas UI. You can click Add Data to show a dropdown list of either importing a file or inserting document. You can see Insert Document in Figure 5-20.

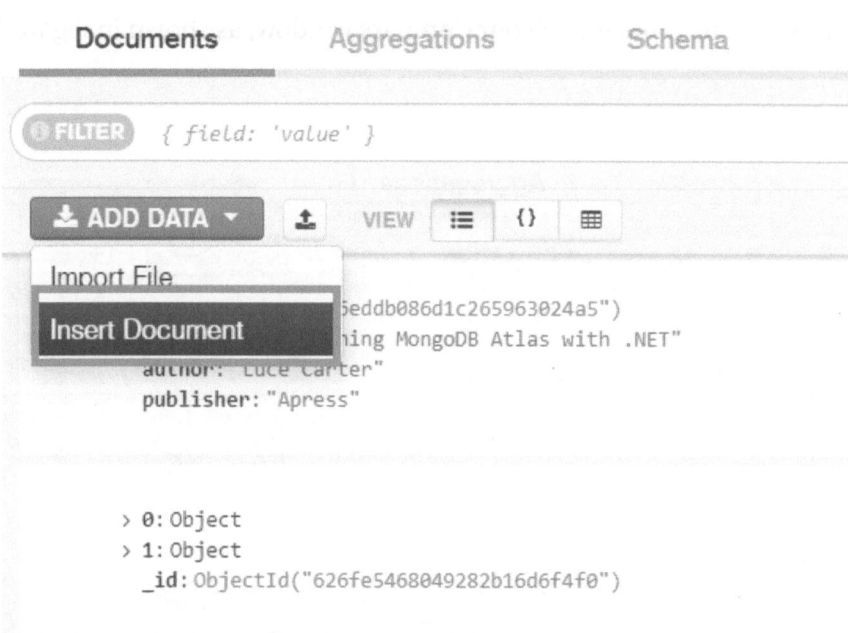

Figure 5-20. *Add data menu for a collection in Compass*

Clicking Insert Document will bring up a dialog box for inserting a document, the same as seen in the Atlas UI.

Hovering your cursor over an existing document inside the UI will bring up a menu on the right, including a pencil icon for editing the document, the same as in Atlas.

This same menu also allows you to delete an existing document from your collection.

Aggregation Builder

In Chapter 2, you learned about aggregations, a way to do more complex queries and manipulations of your data.

Compass has an aggregation builder tool, which allows you to add stages to your aggregation pipeline. You can filter or change data at each stage and see a sample output at each of those stages, to help you visually build your pipeline. It can be accessed from the Aggregation tab, near the top of your Compass window, as shown in Figure 5-21.

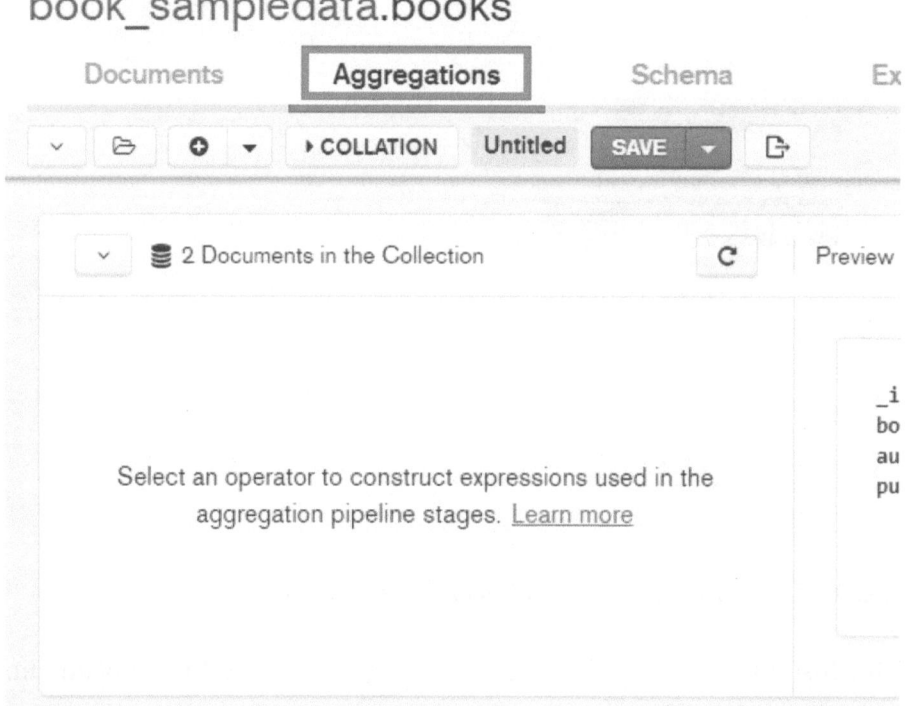

Figure 5-21. *Aggregations tab available inside Compass*

You can then start to add stages and build out queries for those stages, seeing the output at each stage, as shown in Figure 5-22.

Figure 5-22. *Aggregation Pipeline Builder inside Compass*

This is a fun aggregation being run against the movies collection inside the
`sample_mflix` database. It starts by carrying out a `$group` stage to get the average IMDb
rating for each year with data inside our collection and saving it inside a new field called
"avgRating".

As you can see, the outputted rating has too many decimal places and looks a bit
messy. A `$project` stage is then carried out on that output from the previous stage, to
carry out rounding on our new "avgRating" field to two decimal places, saving it in a new
field called "averageRating".

What makes this builder so powerful, though, is the ability to share it in different
ways. Save allows the query to be saved so it can be loaded in future, should you want to
access the pipeline again.

Even better than that is the ability to export the pipeline to your language of choice!
You can do this by clicking the export button along the top, as shown in Figure 5-23.

Figure 5-23. *Export pipeline to code button inside Compass Aggregation Builder*

From the resulting window, you can then pick your language of choice, such as C#, from the Export Pipeline To dropdown box, and it will generate valid pipeline code you can use in your language. The average rating pipeline can be seen as C# code in Figure 5-24.

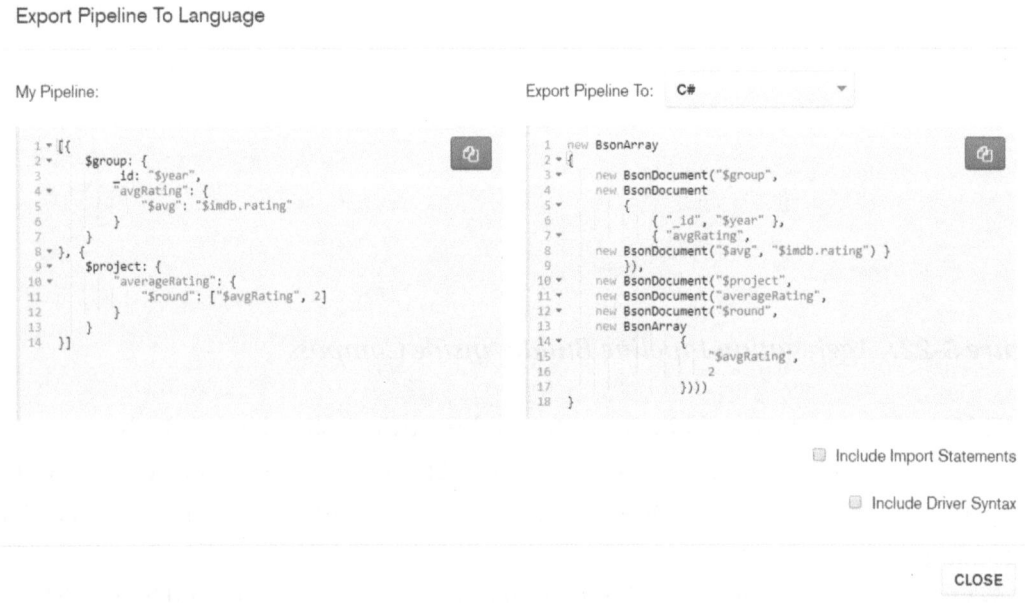

Figure 5-24. *Pipeline to fetch average rating data, available as C# code*

One cool thing to note as well is that you can export to languages such as C#, Python, Java, etc., or if you select Node in the dropdown box, it generates code as an object inside an array, the same as the MongoDB Shell syntax. This allows it to be shared with others via an import in Compass, by using the down arrow next to the + button inside the Aggregation window and selecting New Pipeline from Plain Text, as seen in Figure 5-25.

Figure 5-25. *Creating a new pipeline from text inside Compass*

Visual Studio Code Extension

One of the newer tools for accessing and manipulating your data in Atlas is an extension for Visual Studio Code (VS Code).

It can be downloaded from the VS Code Extension Marketplace by searching for MongoDB and installing the verified official extension, as shown in Figure 5-26.

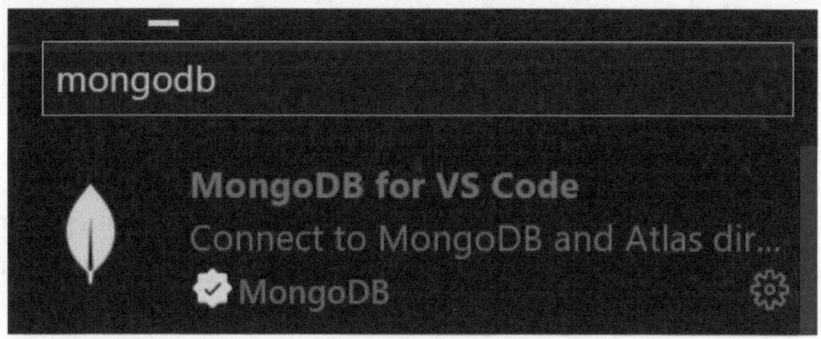

Figure 5-26. *Official MongoDB Extension in the VS Code Extension Marketplace*

Once installed, you can access it from the Command Palette or by selecting the appropriate icon from the Activity Bar. You can then connect to your cluster, using the same connection string as used for Compass, and carry out the usual CRUD operations against the collections.

Figure 5-27 shows a connected extension with the databases and their collections available for browsing.

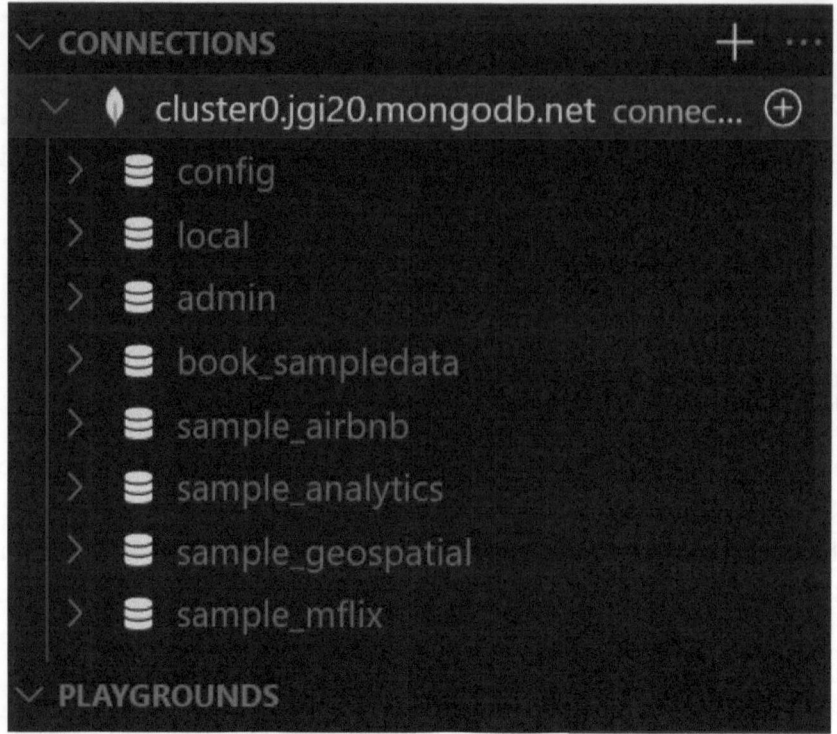

Figure 5-27. *VS Code MongoDB Extension connected to an Atlas cluster*

The VS Code Extension has a built-in playground as well, so you can prototype your aggregations and queries before adding them to your code. It even provides syntax highlighting, autocomplete, and stage snippets for your aggregations.

When you create a playground for the first time, it even gives you a bunch of sample code, against a database and collection it will create for you, to help you understand how to interact with your data, from right within VS Code, as shown in Figure 5-28.

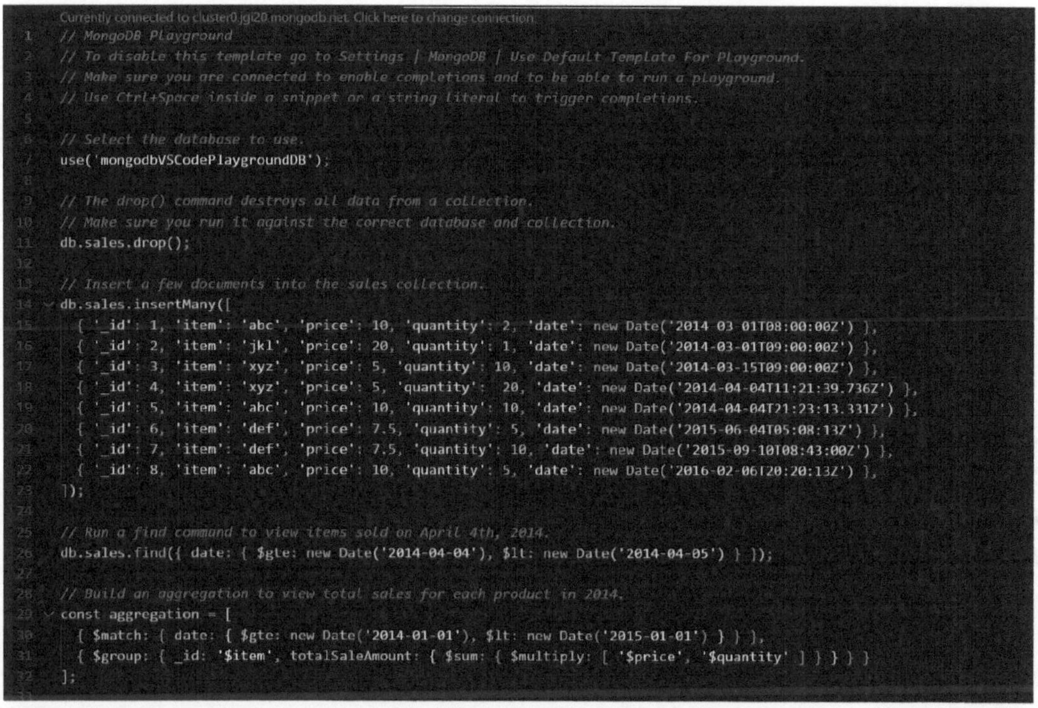

Figure 5-28. *Default New Playground with example code*

Reading Data

Browsing data in the extension is also very easy. You can click into the databases and their collections from the same area as shown in Figure 5-27, and it will allow you to see all documents in that collection. You can then click one and open it up, as seen in Figure 5-29.

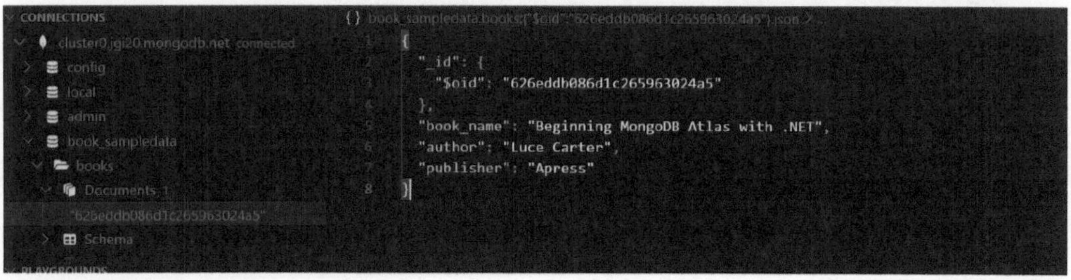

Figure 5-29. *Reading an existing document with VS Code MongoDB Extension*

Creating Data

The playground that comes with the extension allows you to use the syntax you already know, from the MongoDB API Language used in MongoDB Shell, to easily manipulate your data.

Figure 5-30 shows how to use `insertMany` to add multiple documents into the books collection.

```
// Insert a few documents into the sales collection.
db.books.insertMany([
    { 'book_name': 'Xamarin in Action', 'author': 'Jim Bennett', 'publisher': 'Manning' },
    { 'book_name': 'Xamarin.Forms Essentials', 'author': 'Gerald Versluis', 'publisher': 'Apress' },
]);
```

Figure 5-30. *Using the insertMany function inside a playground in VS Code MongoDB Extension*

After clicking the run button in the top right of the window to execute your playground code, you will be met with a pane similar to Figure 5-31, showing the object ids of the successfully inserted documents.

```
Include Import Statements | Include Driver Syntax
1   {
2       "acknowledged": true,
3       "insertedIds": {
4         "0": {
5           "$oid": "62f90f0e42836088559f1a59"
6         },
7         "1": {
8           "$oid": "62f90f0e42836088559f1a5a"
9         }
10      }
11  }
```

Figure 5-31. *Inserted document ids from a successful playground execution*

Updating Data

Your data can be updated using the same helper methods you would use from the MongoDB Shell. Figure 5-32 demonstrates how you can update an existing document using updateOne. The filter criteria are passed as the first parameter, followed by a document describing the change.

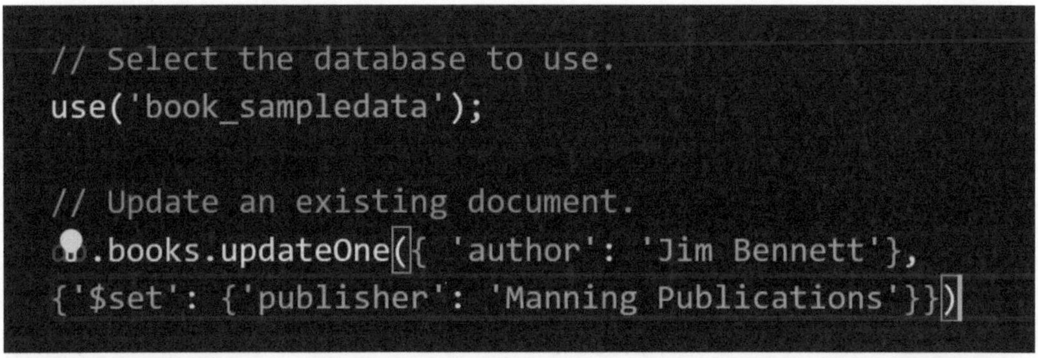

```
// Select the database to use.
use('book_sampledata');

// Update an existing document.
db.books.updateOne({ 'author': 'Jim Bennett'},
{'$set': {'publisher': 'Manning Publications'}})
```

Figure 5-32. *Calling updateOne to update an existing document from the VS Code Playground*

Deleting Data

As was the case when creating or updating data, deleting data can be done using the same helpers as were used with the MongoDB Shell. Figure 5-33 shows you how to delete an existing document using the deleteOne function, passing in a filter to help locate the document for deletion.

```
// Select the database to use.
use('book_sampledata');

// Update an existing document.
db.books.deleteOne({ 'author': 'Jim Bennett'})
```

Figure 5-33. *Calling deleteOne to remove a document in the VS Code playground*

In addition to being able to manipulate your data directly from the VS Code UI, the VS Code extension gives you access to the shell from the command palette.

Data API

Atlas App Services provides a feature known as the Data API, which allows you to access your data over secure and fully managed HTTPS calls. This is good for situations where HTTPS calls are enough, or the only way to communicate with your cluster, such as prototyping, functions on the edge, or IoT devices.

This feature can be enabled on your cluster by going to Data API from the left side of the Atlas UI, as shown in Figure 5-34.

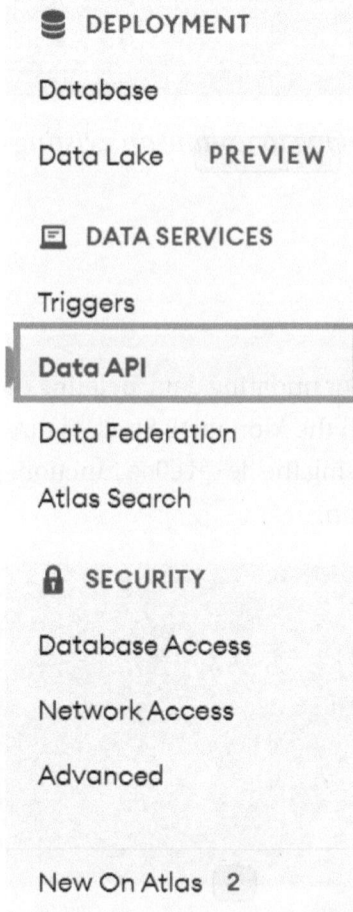

Figure 5-34. Data API from the left-side menu in Atlas

You will then be prompted to select the cluster you want to enable it on, at which point you simply tick the box and click Enable Data API.

This will then bring you to a screen similar to Figure 5-35. You will want to click the dropdown next to the cluster you selected to change the access rules, such as to Read and Write.

Figure 5-35. *Changing Data API Access for a cluster in Atlas UI*

Carrying out requests to the Data API endpoint requires an API key to prove you have authorization to access the cluster. There is a Create API Key button in the top right of the Atlas UI, allowing you to create a new API key, as shown in Figure 5-36.

Figure 5-36. *Create API Key inside Data API in Atlas*

Reading Data

Once you create an API key, it will display a modal window with your new API key, as well as some sample code, by default showing how to POST with `curl` to your new Data API endpoint, allowing you to view data.

Figure 5-37 shows the dropdown box, allowing you to choose which language you want sample code for, to test your Data API, including C#.

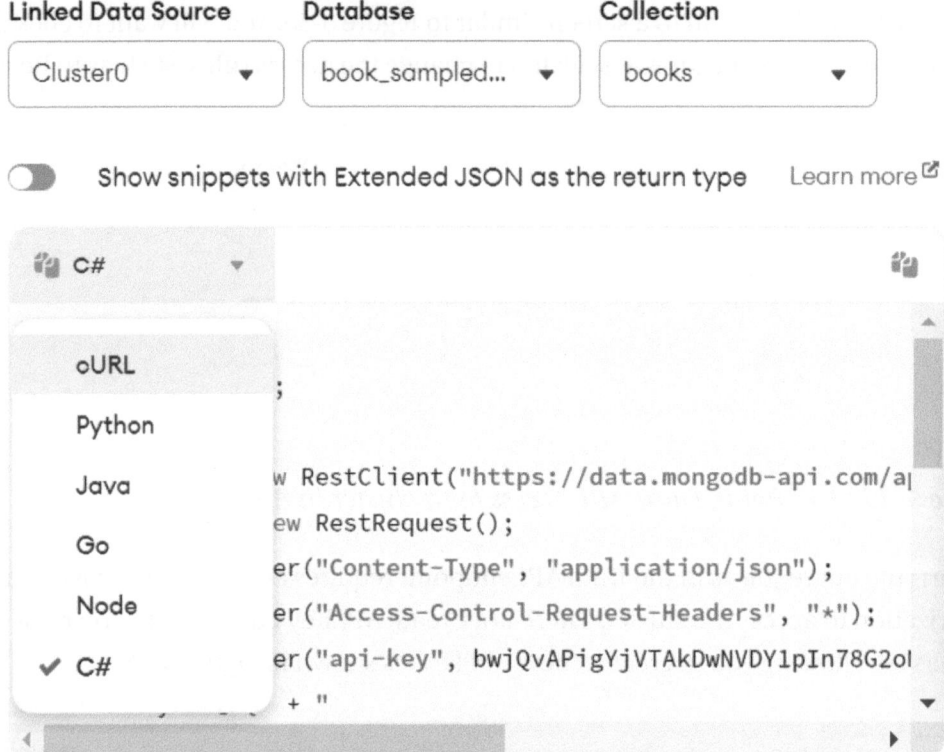

Figure 5-37. *Selecting language in the Data API test window*

Reading the data can be as simple as running the `curl` command using the code in the Test Data API window, as seen in Figure 5-38.

```
PS C:\Users\LuceCarter> curl --location --request POST 'https://data.mongodb-api.com/app/data-hrdlp/endpoint/data/v1/action/findOne' --h
quest-Headers: *' --header 'api-key:                                                                    --data-raw '{
>>     "collection":"books",
>>     "database":"book_sampledata",
>>     "dataSource":"Cluster0"
>> }'
>>
{"document":{"_id":"626eddb086d1c265963024a5","book_name":"Beginning MongoDB Atlas with .NET","author":"Luce Carter","publisher":"Apress
```

Figure 5-38. *Running a curl command inside terminal*

Creating Data

Creating data with the Data API is also easy. You can easily call the insertOne endpoint, passing the document you wish to insert, as shown in Figure 5-39.

```
PS C:\Users\LuceCarter> curl --request POST 'https://data.mongodb-api.com/app/data-hrdlp/endpoint/data/v1/action/insertOne'
                                              ' --data-raw '{
>>      "dataSource": "Cluster0",
>>      "database": "book_sampledata",
>>      "collection": "books",
>>      "document": {
>>        "name": "Future Book on DataAPI",
>>        "author": "TBC",
>>        "publisher": "Apress"
>>      }
>>  }'
{"insertedId":"62f92c1939113b0241ef5817"}
PS C:\Users\LuceCarter>
```

Figure 5-39. *Inserting a document using the Data API from the terminal*

Updating Data

The updateOne endpoint can be used to update an existing document. You can see this in action in Figure 5-40. A filter is passed first to help identify what we want to modify, followed by a document containing our updated fields and values.

```
PS C:\Users\LuceCarter> curl --request POST 'https://data.mongodb-api.com/app/data-hrdlp/endpoint/data/v1/action/updateOne'
                                              --data-raw '{
>>      "dataSource": "Cluster0",
>>      "database": "book_sampledata",
>>      "collection": "books",
>>      "filter": { "author": "Luce Carter" },
>>      "update": {
>>        "name": "Beginning MongoDB Atlas with .NET",
>>        "author": "Luce Carter",
>>        "publisher": "Apress"
>>      }
>>  }'
{"matchedCount":1,"modifiedCount":1}
```

Figure 5-40. *Updating an existing document using the Data API from the terminal*

Deleting Data

Finally, we come to delete. As you can guess, we can call the deleteOne endpoint and pass a filter to identify the document to be deleted.

Figure 5-41 shows this in action.

```
PS C:\Users\LuceCarter>  curl --request POST 'https://data.mongodb-api.com/app/data-hrdlp/endpoint/data/v1/action/deleteOne'
                                              ' --data-raw '{
>>      "dataSource": "Cluster0",
>>      "database": "book_sampledata",
>>      "collection": "books",
>>      "filter": { "author": "Gerald Versluis" }
>>  }'
{"deletedCount":1}
PS C:\Users\LuceCarter>
```

Figure 5-41. *Deleting an existing document using the Data API from the terminal*

Drivers

When developing new applications, regardless of programming language, drivers are often used to provide additional capabilities. Although you can interact with your data via the new Atlas Data API as seen previously, when developing an application the best way to interact with your MongoDB Atlas cluster or databases is to utilize a driver.

MongoDB currently supports the following official drivers:

- C
- C++
- C#
- Go
- Java
- Node.js
- PHP
- Python
- Ruby
- Rust
- Scala
- Swift

There are also a bunch of community support libraries for a variety of languages including R, Dart, and Perl.

The best part of the official drivers is that they are actively maintained by engineers at MongoDB and updated to include new MongoDB features and receive bug fixes, security patches, and performance improvements.

Later in this book, we will use the official `MongoDB.Driver` NuGet package for C# to interact with a MongoDB Atlas database and do some fun things!

GraphQL

Atlas App Services, as well as offering a way to interact with your data via HTTPS with the Data API, also allows you to use *GraphQL*, a powerful alternative to *REST*.

By creating a Realm application that points to your MongoDB Atlas cluster and setting up some basic requirements such as data access rules and defining schema, you gain access to a GraphQL endpoint.

Figure 5-42 shows the App Services tab inside the Atlas UI in the browser, that you can click to access all the great services within Atlas App Services, including GraphQL.

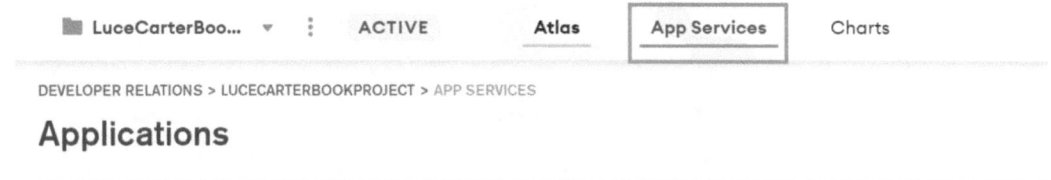

Figure 5-42. *App Services tab inside Atlas UI in the browser*

If you have Data API already enabled, you will find an existing app already inside App Services called "data," and you can click this to load a view as shown in Figure 5-43.

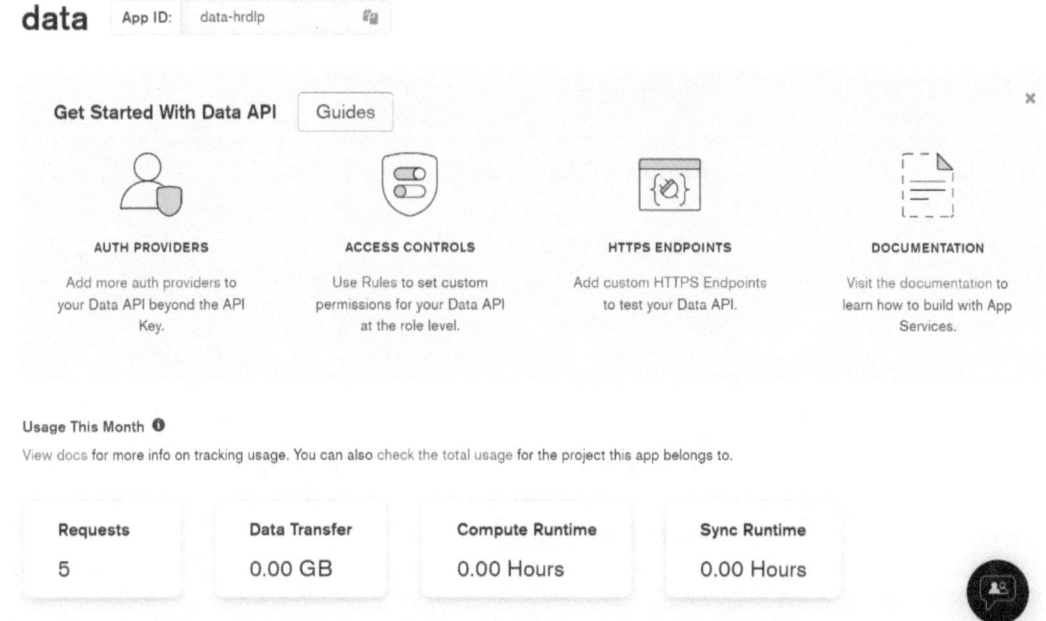

Figure 5-43. *View of an App Services application inside Atlas in the browser*

Although MongoDB has a flexible schema, meaning you don't need to have one defined, you will need a schema defined for using GraphQL. The structure of your data needs to be known so that it can be queried consistently.

The Schema area can be accessed from the left-side menu in Atlas, under the Data Access header, as shown in Figure 5-44.

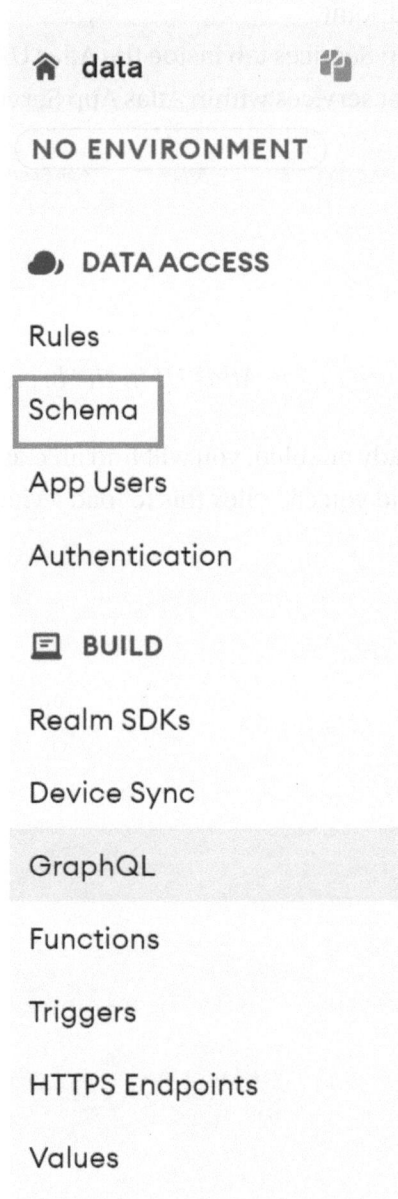

Figure 5-44. *Access Schema inside App Services*

You can then select the collection that you wish to enable GraphQL for, and if it contains data, App Services will be able to generate a schema for you, by clicking the button as shown in Figure 5-45.

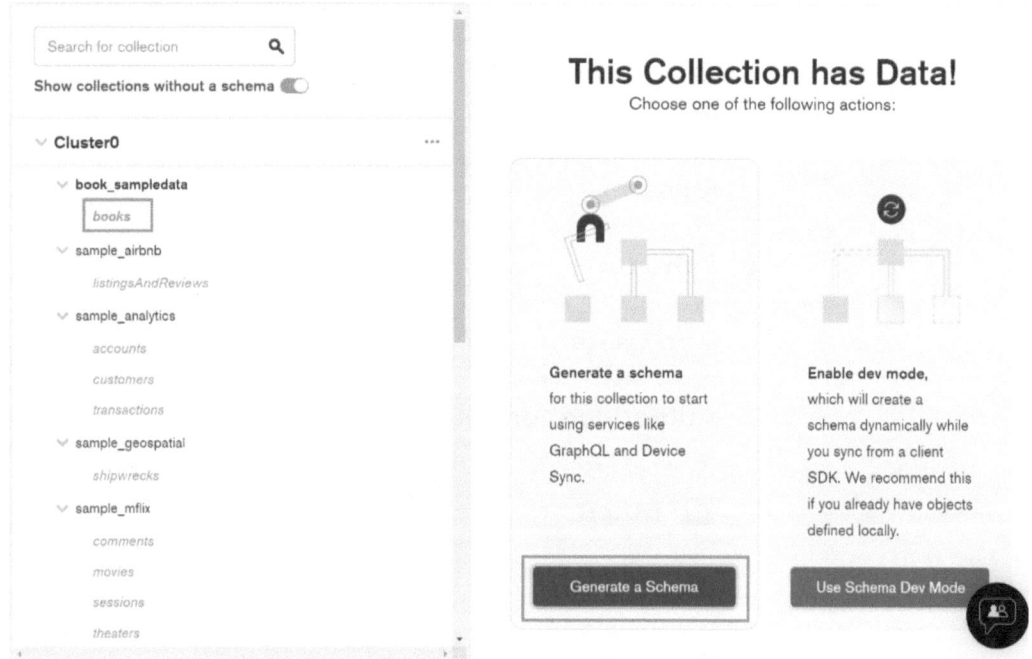

Figure 5-45. *Atlas offering to generate a schema for a collection with data*

Once you generate the schema against your collection and save it, you can access GraphQL from the left, as seen in Figure 5-46.

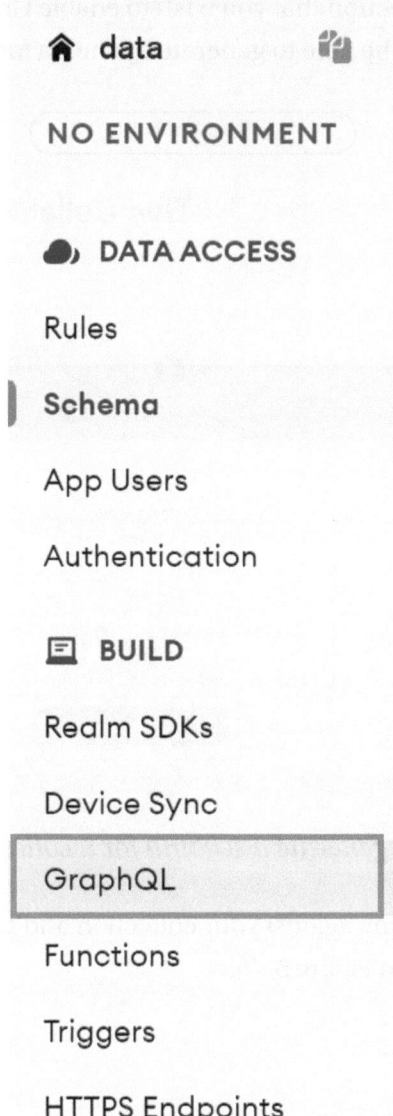

Figure 5-46. *Accessing GraphQL inside Atlas UI in the browser*

Reading Data

When you access GraphQL, you will be met with *GraphiQL*, an interactive GraphQL environment. It will already have a query written to access your data, and clicking the play button will run this query, resulting in documents being returned, as seen in Figure 5-47.

```
GraphiQL    ▶    Prettify    Merge    Copy    History

 9   #
10   # Queries typically start with a "{" character. I      ▼ {
11   # with a # are ignored.                                    "data": {
12   #|                                                   ▼      "book": {
13   # You can use the Documentation Explorer to the r              "_id": "626eddb086d1c265963024a5",
14   # are available for you to query for your applica             "author": "Luce Carter",
15   #                                                             "name": "Beginning MongoDB Atlas with .NET",
16   # If nothing is appearing in the Documentation E:             "publisher": "Apress"
17   # to resolve some warnings or errors before usin           }
18   # for the Book type has been generated for you be        }
19   #                                                    }
20   #
21   # Keyboard shortcuts:
22   #
23   #   Prettify Query:  Shift-Ctrl-P (or press the pr
24   #
25   #     Merge Query:  Shift-Ctrl-M (or press the me
26   #
27   #       Run Query:  Ctrl-Enter (or press the play
28   #
29   #   Auto Complete:  Ctrl-Space (or just start typ
30   #
31
32 ▼ query {
33 ▼   book {
34       _id
35       author
36       name
37       publisher
38     }
39   }
```

Figure 5-47. GraphQL query and result in App Services

Creating, Updating, and Deleting Data

GraphQL uses something called *mutations* when it comes to manipulating your data. App Services GraphQL already has mutation functions available for all the standard MongoDB operations including insertOne, insertMany, updateOne, updateMany, deleteOne, and deleteMany.

Figure 5-48 shows a mutation, using the existing insertOneBook it has generated for my collection based on the schema, adding a new book. It takes a data argument for the document we want to insert.

```
1   mutation
2 ▾ {
3 ▾   insertOneBook(data: {
4       name:"Xamarin.Forms Essentials",
5       author: "Gerald Versluis",
6       publisher: "Apress"
7     })
8   }
```

Figure 5-48. *insertOne mutation inside GraphQL*

Updating and deleting data is also very similar. Figure 5-49 shows both mutations, as more than one can be run at once inside GraphiQL.

```
1   mutation
2 ▾ {
3     updateOneBook(query: {
4       author: "Gerald Versluis"
5     },
6     set: {
7       publisher: "APress"
8     })
9
10    deleteOneBook(query:{
11      author: "Gerald Versluis"
12    })
13  }
```

Figure 5-49. *Update and delete mutations inside GraphiQL*

The functions being called in Figure 5-49 might seem quite intuitive as they follow a similar coding pattern to other tools, such as MongoDB Shell.

The *updateOneBook* function takes two parameters: a query, to help identify the document to update, and a set, passing the value to be changed. You only need to pass the field to be changed and not every field, due to the power of GraphQL.

deleteOneBook takes a query to identify the document to be deleted.

Summary

- MongoDB Atlas has a lot of tools for accessing and manipulating data.

- What tool you choose comes down to personal preference and use case.

- You can use Compass, not only as a powerful GUI but to help you visually build aggregation pipelines.

- As we will use in later chapters, MongoDB supports drivers in many languages, including C#.

Wow, what a bumper chapter!

We have learned all the amazing ways there are to access and change your data, all depending on your preference. This is also the end of Part 2.

In Part 3, we will build an application with .NET 6 and MongoDB Atlas. In Chapter 6, we will look at creating a new project that we can build upon throughout Part 3.

PART III

Building a Project

Creating the Application

How exciting! We know all about MongoDB's developer data platform now and even all the ways you can interact with your data. Over the course of the next few chapters, we are going to build out an application, using C# and the MongoDB driver, and begin to carry out the most common operations: create, read, update, and delete.

The plan over the next few chapters is to build a minimal Web API, using ASP.NET Core with .NET 6.0. We are creating a Web API project as we don't need to have a front end in order to access this data. As with many enterprise applications, the Web API can live as a standalone project, with endpoints for each CRUD operation, that can be accessed from any other project, including ones in other languages and frameworks, or from something like a ASP.NET Core MVC app that you may choose to create later.

I am a big fan of games, whether they are video games, card games, or board games. So, our application is going to allow us to store games in our database and include information in the document, such as what type of game it is, the number of players required, and the maximum number of people who can play. Whether you like games too or not, this application will share the fundamental knowledge required to go on and store whatever data you like!

In this chapter, we are going to look at the options for how to create a Web API project. By the end of the chapter, we will have an application created and ready to be built upon as we go through the later chapters in Part 3.

If you wish to skip the creation part and instead want to start from an existing codebase, you can clone a git repository found at `https://github.com/LuceCarter/beginning_mongodb_atlas_dotnet` on a branch called "start."

© Luce Carter 2024
L. Carter, *Beginning MongoDB Atlas with .NET*, https://doi.org/10.1007/978-1-4842-9550-2_6

Tooling

The great thing about .NET development these days is the abundance of choice around tooling. You can choose between creating and managing your projects with a command-line interface (CLI) and editing inside a powerful text editor such as VS Code, or use an integrated development environment (IDE) to create, edit, and run your applications. Which tool you use is up to you. The following sections will show you how to create a new ASP.NET Core Web API project in the relevant tools. Follow whichever section you prefer – the end result is always the same.

.NET SDK – CLI

Currently, when you download the .NET SDK, whether that is .NET 6 or some earlier versions, not only do you get the SDK for creating and running your applications, but you also get a CLI tool that you can call with dotnet.

This tool allows you to create projects, manage NuGet packages, build, run, and much more.

Out of the box, when you run dotnet new, you will be met with a bunch of common templates that you can use to create your project. An example of this is shown in Figure 6-1.

```
 ■ Dev > dotnet new
The 'dotnet new' command creates a .NET project based on a template.

Common templates are:
Template Name            Short Name       Language        Tags
----------------------   --------------   ------------    ----------------------
ASP.NET Core Web App     webapp,razor     [C#]            Web/MVC/Razor Pages
Blazor Server App        blazorserver     [C#]            Web/Blazor
Class Library            classlib         [C#],F#,VB      Common/Library
Console App              console          [C#],F#,VB      Common/Console
Windows Forms App        winforms         [C#],VB         Common/WinForms
WPF Application          wpf              [C#],VB         Common/WPF

An example would be:
   dotnet new console

Display template options with:
   dotnet new console -h
Display all installed templates with:
   dotnet new --list
Display templates available on NuGet.org with:
   dotnet new web --search
```

Figure 6-1. *dotnet new common default templates*

To create our Web API project, run the following code, substituting the part after -o with what you choose to call your project:

```
dotnet new webapi -o <name of project>.
```

By calling -o and the name of the project, it will create a folder with the name of your application and put the files inside there. If you run it without -o and the name, it will create an application in the location where you ran the command, with the application named after the parent folder. So, it's always easier to add -o and a name, as shown in Figure 6-2.

```
 ■ Dev > dotnet new webapi -o mongodb-dotnet-webapi
The template "ASP.NET Core Web API" was created successfully.

Processing post-creation actions ...
Running 'dotnet restore' on C:\Users\LuceCarter\Dev\mongodb-dotnet-webapi\mongodb-dotnet-webapi.csproj ...
   Determining projects to restore ...
   Restored C:\Users\LuceCarter\Dev\mongodb-dotnet-webapi\mongodb-dotnet-webapi.csproj (in 387 ms).
Restore succeeded.
```

Figure 6-2. *Creating a new Web API application with the dotnet CLI*

This will create a bunch of boilerplate code, as seen in Figure 6-3. This includes the code required to run the application, and some example code, provided to help you get started, such as WeatherForecast.cs.

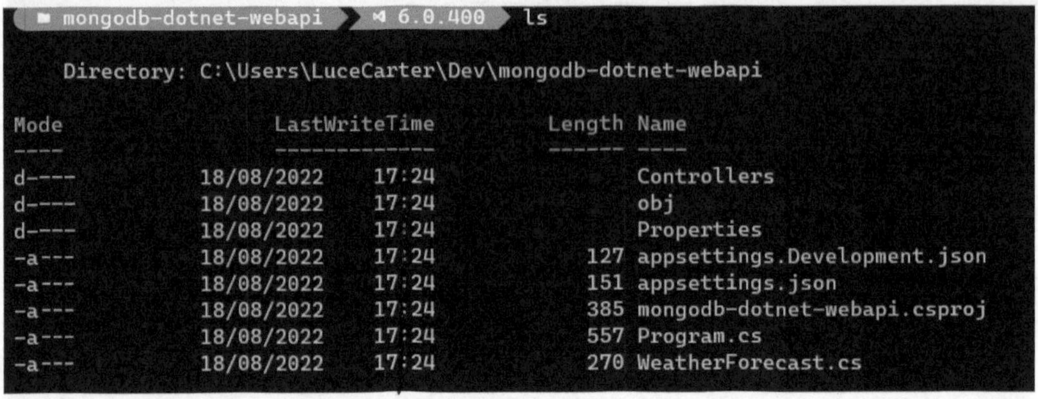

Figure 6-3. *Default project files and folders from dotnet new Web API*

To test everything is set up correctly, run the command dotnet run from inside the project folder. It will then build and run the application, giving you an output similar to Figure 6-4.

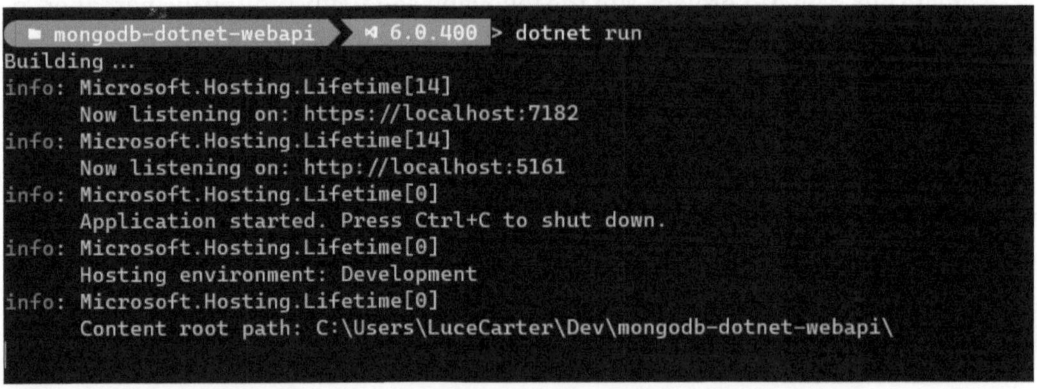

Figure 6-4. *Successful build and deploy with dotnet run command*

You can then visit the URL that it provided you in the terminal, in this case http://localhost:5651, although your port number may be different. Of course, because this is a Web API, loading the root will simply return nothing because there is no front end for the server to deliver. However, all ASP.NET Core Web API projects come with Swagger, a document builder for your endpoints. So, if you append `http://localhost:5651/` `swagger/index.html` to your localhost URL, `swagger/index.html` for example, you will be met with a screen similar to Figure 6-5.

Figure 6-5. *Swagger running on the default ASP.NET Core Web API template code*

This is all we need for now. At the end of this chapter, we will tidy up the code a little bit to remove the files we don't need going forward.

Integrated Development Environment (IDE)

With .NET and C# development, we are spoiled with two great IDEs: Microsoft Visual Studio and JetBrains Rider.

Although Visual Studio has been around for decades and is the flagship IDE from Microsoft, in recent years, JetBrains has developed an IDE called Rider, targeted at .NET developers, which has grown in popularity.

Which you choose is entirely down to personal preference. So, in this section, we will cover how to create a ASP.NET Core Web API project in both tools.

Visual Studio 2022

Visual Studio 2022 is the latest version from Microsoft at the time of writing. There is even a free version, community edition, which can be used to follow along with this book. Of course, you may use a different version, like 2019 or even another version that has come out when you are reading this, so screenshots may vary slightly.

Create a new Web API project in Visual Studio by selecting File ➤ New ➤ Project and searching for Web API, or selecting WebAPI in the Project Type dropdown box and selecting ASP.NET Core Web API, as seen in Figure 6-6.

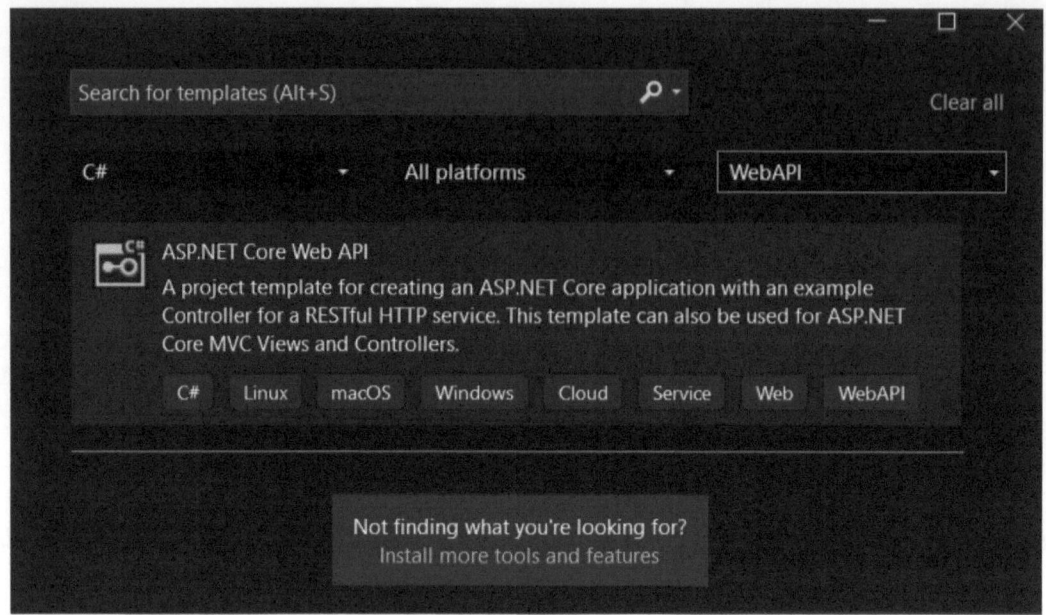

Figure 6-6. *Finding the ASP.NET Core Web API project type in Visual Studio new project wizard*

To finish up, click through the wizard, giving your project a name, leaving the additional information settings as their default, and creating the project.

If you then run this application from within Visual Studio, it will automatically open a window in your default browser and navigate to the Swagger page, resulting in the same view as seen in Figure 6-5.

We are now set up within Visual Studio 2022 to start building our application. At the end of the chapter, we will remove some of the boilerplate code that comes with the template that we don't need.

JetBrains Rider

As mentioned earlier in the chapter, Rider from JetBrains is proving to be a popular tool among many developers. For this reason, it deserves a section in this book, in the event you would like to use it.

When you open Rider, it brings up a Welcome to JetBrains Rider window which will show recent projects (if you have opened any) and other options. At the top, there is a New Solution button, which is what we want, as we want to create a new project. You can see this button in Figure 6-7.

Figure 6-7. *New Solution button inside the welcome window in JetBrains Rider*

Clicking this button will then show the New Solution window inside Rider.

Select ASP.NET Core Web Application from the left-hand side menu, under the .NET/.NET Core heading.

This then updates the main part of the window with details you can fill out, such as the solution name, solution directory, type, language, and other information. To create a Web API, we want to select Web API from the Type dropdown box, as seen in Figure 6-8.

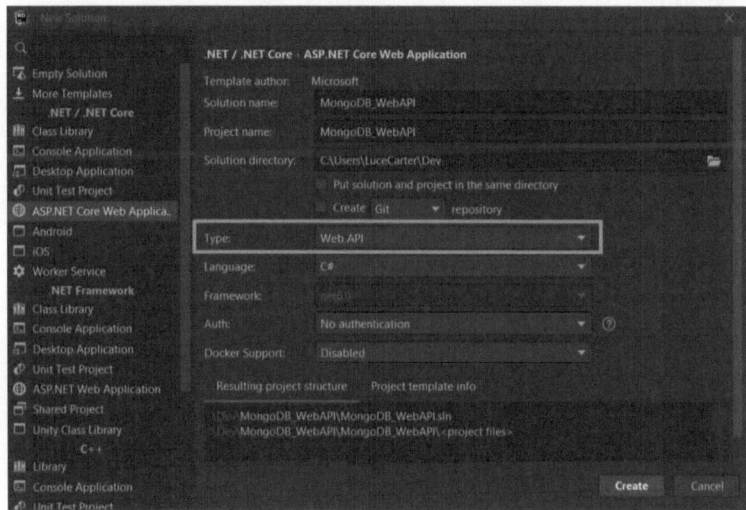

Figure 6-8. *Selecting ASP.NET Core Web API project in Rider*

Once created, this project generates the same boilerplate code as you get from the .NET CLI and Visual Studio.

If you run it, it again opens your default browser in a new window, at the Swagger endpoint, showing the default GetWeatherForecast endpoint that comes with the boilerplate code. You can see how this would look in Figure 6-5.

Cleaning Up

Regardless of which option you selected for creating the project, we now have a ASP.NET Core Web API project ready. However, they all came with some boilerplate code that we don't need in our application, so we are going to clean up in the following steps:

1. Delete *WeatherForecast.cs* from the root of the project. This is a model class for the weather forecast that we won't need.

2. Inside the Controllers folder, delete *WeatherForecastController.cs*. Controllers in ASP.NET Core are essentially handlers for requests and provide endpoints for the application. We will go ahead and create our own in a future chapter.

3. Run the application to ensure it still builds and runs. It should load the Swagger page again, but this time show something similar to Figure 6-9, which just means that we don't have any endpoints defined for Swagger to document.

MongoDB_ASPNetCore_WebAPI 1.0 OAS3
https://localhost:7133/swagger/v1/swagger.json

No operations defined in spec!

Figure 6-9. *Swagger page when no controllers present in project*

Summary

- Projects can be created in multiple ways: CLI or with IDEs such as Visual Studio and Rider.

- What you use is down to personal preference.

- The template for a .NET Core Web API project comes with useful boilerplate code to get up and running with straight away.

Yay! We have our foundation project ready to go that we can start to build upon. Join us in the next chapter as we add the MongoDB NuGet Package, store some configuration, and connect to our Atlas cluster.

CHAPTER 7

Adding MongoDB

It's time for us to bring together two amazing developer technologies: .NET and MongoDB. By the end of this chapter, we'll be able to connect our application to our Atlas cluster and return a list of existing databases to test the connection. This gives us a great starting point to build upon in later chapters.

Going forward, I am going to be using screenshots from Visual Studio, but you can work on your application however you want.

Add the MongoDB NuGet Package

The first step is to add the MongoDB C# driver. This gives us access to methods we can use to interact with our Atlas database and collections.

1. Open the NuGet Package Manager (from the Project menu) and search in the Browse tab for *MongoDB*.

2. A few options will appear. We want *MongoDB.Driver*, as seen in Figure 7-1.

3. Install the latest stable version.

4. Accept any license agreements that appear.

5. Close NuGet Package Manager.

© Luce Carter 2024
L. Carter, *Beginning MongoDB Atlas with .NET*, https://doi.org/10.1007/978-1-4842-9550-2_7

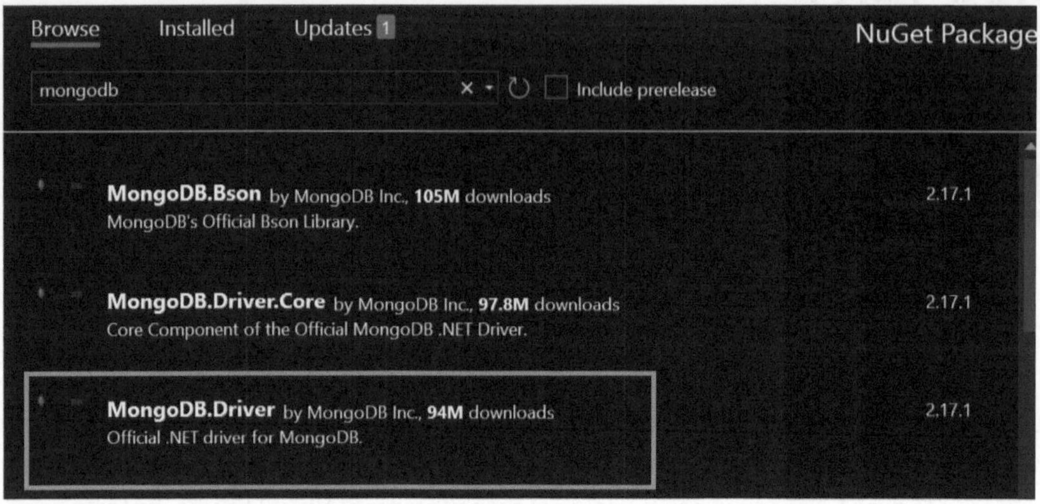

Figure 7-1. *The MongoDB.Driver inside NuGet Package Manager*

For those of you who prefer to use the Package Manager Console, the following line will install the driver:

```
Install-Package MongoDB.Driver
```

Store Connection String

The next step is to store our connection string. We will use this connection string when we create the MongoDB client object later, so it knows how to connect to our Atlas cluster.

Fetch Your Connection String from Atlas

Let's first look at how to get our connection string from Atlas, which we can then add into our application.

1. Open Atlas and log in so you are taken to the database page.

2. Click the Connect button along the top, as shown in Figure 7-2.

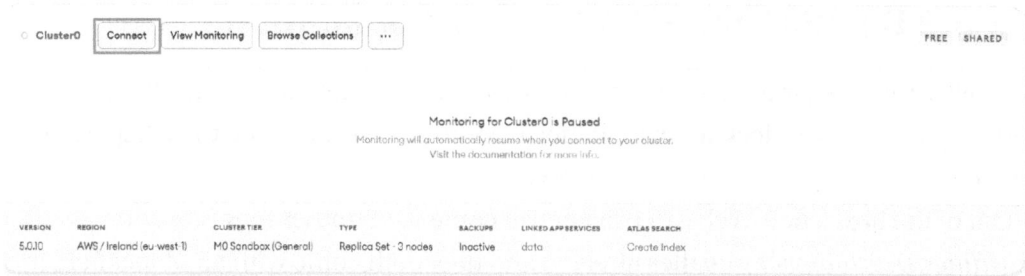

Figure 7-2. *Connect to cluster button in the Atlas UI*

3. Select Connect your application.

4. Select Drivers as the connection method.

5. Select C#/.NET from the Driver dropdown and copy the provided connection string, as shown in Figure 7-3.

Connect to Cluster0

✔ Setup connection security ❯ ✔ Choose a connection method ❯ **Connect**

1 Select your driver and version

DRIVER	VERSION
C# / .NET ▼	2.13 or later ▼

2 Add your connection string into your application code

☐ Include full driver code example

```
mongodb+srv://<username>:<password>@cluster0.jgi20.mongodb.net/?
retryWrites=true&w=majority
```

Replace **<password>** with the password for the **<username>** user. Ensure any option params are URL encoded.

Having trouble connecting? View our troubleshooting documentation

[Go Back]   [Close]

Figure 7-3. *C# connection string provided inside the Atlas UI*

Add the Connection String to Our Application

In ASP.NET Core, regardless of whether it's a Web API, Model View Controller (MVC), or other type of web project, it's a good idea to store strings that need to be kept safe or accessed in multiple locations inside appsettings.

Out of the box, a new Web API project will come with *appsettings.json* and *appsettings.Development.json* files already, and they both come with an object field in there that stores details about logging. These files are used at runtime at different times, depending on how the application is being run – production or development. We will make the changes in both files.

We don't want to remove any existing settings inside these files. Instead, we want to add to it. We will add some configuration later on, so we are going to create our own object field inside the file.

1. Inside *appsettings.json*, add the following as a field to your file (the final result can be seen in Figure 7-4):

```
"GamesDatabaseSettings": {
    "ConnectionString": "<Your Connection String from Atlas UI>"
},
```

```
{
  "GamesDatabaseSettings": {
    "ConnectionString": "<Your Connection String from Atlas UI>"
  },
  "Logging": {
    "LogLevel": {
      "Default": "Information",
      "Microsoft.AspNetCore": "Warning"
    }
  },
  "AllowedHosts": "*"
}
```

Figure 7-4. *Our appsettings.json file with our new GamesDatabaseSettings object*

2. Replace the connection string value between the double quotes with your own connection string, making sure to update it with your username and password.

3. Add the same in the *appsettings.Development.json* file. The end result should look the same as Figure 7-3, just without the *AllowedHosts* field.

Create a MongoDB Client

Now that we have the NuGet package installed and our connection string stored, it's time to put it to use.

We want to create a MongoDB client, which allows us to access a bunch of methods to interact with our cluster.

1. Open *Program.cs* and at the top of the file, add the following two using statements:

```
using MongoDB.Driver;
using MongoDB.Bson;
```

2. Because of the new minimal APIs, reducing the amount of common code needed to be declared, there is no set method in the file, so we want to add the following after the builder has been created:

```
var connectionString = builder.Configuration["GamesDatabase
Settings:ConnectionString"];

MongoClient mongoDBClient = new MongoClient(connectionString);
```

Just like that, we have a MongoDB client created that can talk to our Atlas cluster.

Retrieving a List of Databases

Before we go ahead and start interacting with our collections and documents, it's good to test the connection is working with our MongoDB client we just created. We are going to do that by getting the application to fetch a list of all databases and print out their names.

1. Add the following after the last line, creating the client:

```
List<string> databases = mongoDBClient.ListDatabase
Names().ToList();
```

2. Now we have the list of database names, we can add the following foreach loop to print out the names:

```
foreach(string db in databases)
{
    Console.WriteLine(db);

}
```

3. Run the application, and it will open both a browser window to the Swagger page and also a terminal window, showing the list of database names. It should look similar to Figure 7-5.

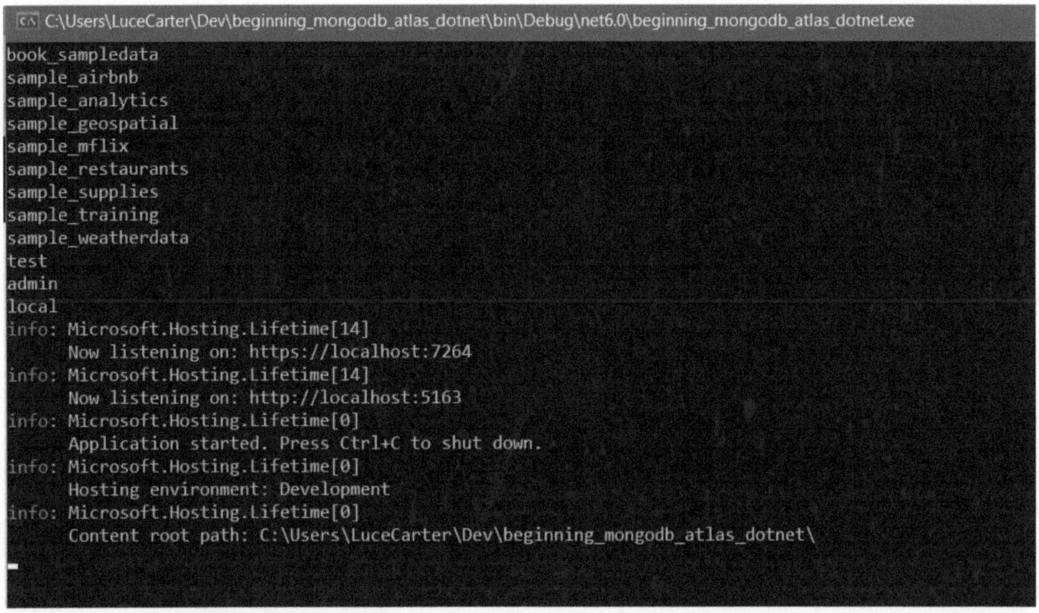

```
C:\Users\LuceCarter\Dev\beginning_mongodb_atlas_dotnet\bin\Debug\net6.0\beginning_mongodb_atlas_dotnet.exe
book_sampledata
sample_airbnb
sample_analytics
sample_geospatial
sample_mflix
sample_restaurants
sample_supplies
sample_training
sample_weatherdata
test
admin
local
info: Microsoft.Hosting.Lifetime[14]
      Now listening on: https://localhost:7264
info: Microsoft.Hosting.Lifetime[14]
      Now listening on: http://localhost:5163
info: Microsoft.Hosting.Lifetime[0]
      Application started. Press Ctrl+C to shut down.
info: Microsoft.Hosting.Lifetime[0]
      Hosting environment: Development
info: Microsoft.Hosting.Lifetime[0]
      Content root path: C:\Users\LuceCarter\Dev\beginning_mongodb_atlas_dotnet\
```

Figure 7-5. *Terminal window showing the result of printing out a list of database names from a MongoDB Atlas cluster*

Final Code

I always find it helpful to have a "Here is what the file should look like" section at the end of any tutorial, because sometimes, it can get confusing where to put things. So, I will do the same in this book to help you. Of course, all code can be found in the GitHub repo, mentioned at the start of the chapter.

If you want to find the code at a certain point in the book, there is a commit of the code at the end of each chapter for easier navigation.

This is what the *program.cs* file should look like now:

```
using MongoDB.Driver;
using MongoDB.Bson;

var builder = WebApplication.CreateBuilder(args);

// Fetch our connection string from appsettings which is available in the
builder's Configuration
var connectionString = builder.Configuration["GamesDatabaseSettings:Connect
ionString"];
```

```csharp
MongoClient mongoDBClient = new MongoClient(connectionString);

List<string> databases = mongoDBClient.ListDatabaseNames().ToList();

foreach(string db in databases)
{
    Console.WriteLine(db);

}

// Add services to the container.

builder.Services.AddControllers();
// Learn more about configuring Swagger/OpenAPI at https://aka.ms/
aspnetcore/swashbuckle
builder.Services.AddEndpointsApiExplorer();
builder.Services.AddSwaggerGen();

var app = builder.Build();

// Configure the HTTP request pipeline.
if (app.Environment.IsDevelopment())
{
    app.UseSwagger();
    app.UseSwaggerUI();
}

app.UseHttpsRedirection();

app.UseAuthorization();

app.MapControllers();

app.Run();
```

Summary

- MongoDB Atlas clusters can be interacted with using the library available as a NuGet package.

- Creating a client to talk to Atlas is as simple as creating the object and passing a connection string.

Just like that, we have stored our connection string in *appsettings*, created a MongoDB client, and shown it's connected successfully by printing out the names of the databases in our cluster.

In the next chapter, we will look at how to create a model of our documents that can be automatically mapped to by the library, a service to use that model to interact with our database and documents using that service, and then insert and interact with them using the driver.

CHAPTER 8

Creating and Interacting with Documents from Code

Things are really starting to come together. We have our Web API project up and running with Swagger so we can see the endpoints. We also have it successfully connecting to our Atlas cluster and listing our databases' names.

In this chapter, we are going to cover quite a few exciting topics. The aim is to refactor the code slightly, so you are able to insert, query, update, and delete documents in your cluster, but be able to handle this data in a way we know well, with classes (a.k.a. models) that represent the fields in our document. On top of this, we are going to create a service that acts as the messenger with our Atlas cluster, using the MongoDB driver we set up in the last chapter.

Adding the Model

The first thing we want to do is create our model that will represent our document locally in code. We will do this using a standard C# class with some additional attributes. As I mentioned in Chapter 6, I love games, so we are going to create a model that will represent our games.

1. Create a folder in the root of your project called *Models*.

2. Create a new C# class inside this folder called *Game.cs*.

© Luce Carter 2024
L. Carter, *Beginning MongoDB Atlas with .NET*, https://doi.org/10.1007/978-1-4842-9550-2_8

3. Add the following using statements to the top of your class file:

```
using MongoDB.Bson.Serialization.Attributes;
using MongoDB.Bson;
using System.ComponentModel.DataAnnotations;
```

4. Add the following inside your new class:

```
[BsonId]
[BsonRepresentation(BsonType.ObjectId)]
public string? Id { get; set; } = String.Empty;

[Required]
public string Name { get; set; }

[Required]
public decimal Price { get; set; }

[Required]
public string Category { get; set; }
```

These additional attributes on top of the usual C# properties come from the MongoDB driver and allow for the automatic serialization and deserialization between C# classes and MongoDB documents, including the handling of different data types.

[BsonId] lets the driver know that this will be the field to use as the _id field in our document, the unique identifier, or primary key.

[BsonRepresentation(BsonType.ObjectId)] is the first instance of how the driver can help map between data types for us. *ObjectId* is the MongoDB data type for its _id field, but this is not a data type that can be used in C#. So, by using this attribute, the supported data type string can be used without needing to handle any type conversions or casting in our applications.

Some fields are also marked as [Required], which does as expected. It means we can decide if a field should always have a value, preventing one being created in code without those values populated.

Creating the Service

Although for an example like this a whole service class might seem excessive, it is useful as your application gets more complex, so we will add one so that you have a solid basis for expanding upon in future.

1. Create a folder in the root of your project called *Services*.

2. Create a new C# class inside this folder called *GamesDatabaseService.cs*.

3. Add the following usings at the top of your class:

   ```
   using BeginningMongoDBAtlasDotNet.Models;
   using MongoDB.Driver;
   ```

   N.B. You may need to adjust the first using statement to match the name of your project.

4. Add `private readonly IMongoCollection<Game> _games;` inside the class. This will hold our list of games.

5. Paste the following into the class, replacing any existing constructor method in the file:

   ```
   public GamesDatabaseService(string connectionString)
           {
                   var mongoDBClient = new MongoClient(connectionString);
                   var database = mongoDBClient.GetDatabase("GamesDB");

                   _games = database.GetCollection<Game>("Games");
           }
   ```

Some of this code will look familiar from Chapter 7 because it is doing the same thing: initializing a connection to our database. However, this time, the connection string is passed to the constructor rather than being hard-coded.

Updating to Use the Service

Before we move on to adding more functionality to our service class for the different create, read, update, and delete methods, we want to update our *Program.cs* file to begin using this new class.

1. Inside *Program.cs*, leave the connectionString variable line in, but remove the other code from Chapter 7 that creates a MongoDB client, fetches a list of databases, and uses a foreach loop to print out all the database names.

The full code for each file will be available at the end of the chapter.

---

2. Add the following line of code after the *// Add services to the container* comment but before any other calls to the builder object:

   ```
   builder.Services.AddSingleton(new GamesDatabaseService
   (connectionString));
   ```

This is the only change we need to make to the file, which is great. But this now means that the service is available to be called from elsewhere. By adding it as a singleton service, it will only be created once. This is ideal as we only need one client instance that talks to Atlas.

Adding Create Methods to the Service

We are now ready to start adding methods to our new games database service, starting with one to create new documents in our database. The MongoDB driver makes this really easy, thankfully. We are going to take advantage of two methods available for creating one document or creating many documents at once.

1. Inside *GamesDatabaseService.cs*, add the line `public void CreateOne(Game game) => _games.InsertOne(game);` after the constructor. This takes a single game object and passes it to the *InsertOne* method. As we learned earlier, the driver takes care of automatically mapping between our C# object and a MongoDB document.

2. Below the previous line, add `public void CreateMany(Game[] games) => _games.InsertMany(games);`. This takes an array of game objects and inserts them, with the driver again able to take care of how to do that.

Adding Read Methods to the Service

So, we have the ability to add new documents, but what about reading existing ones? Well, that's really easy too! We can add two methods here: one for retrieving all game documents from our collection and one for fetching a single document.

1. Below the create methods, add `public List<Game>Get() => _games.Find(game => true).ToList();`, which fetches all documents and converts them to a list. *Find* takes a filter to help identify what you want, so by passing true, we are requesting every document.

2. Below that, add public Game GetOne(string id) => _games. Find(game => game.Id == id).FirstOrDefault();, which uses the id passed in as the filter.

Adding an Update Method to the Service

It's good that we can now create and read documents, but it's also important that we are able to update existing documents.

1. Add `public void UpdateOne(string id, Game updatedGame) => _games.ReplaceOne(game => game.Id == id, updatedGame);` below our previous methods. This takes two parameters this time: the id of the document we want to update and the new updated game object. The *ReplaceOne* method then takes the filter to identify the document to update and then the document to replace it with.

Adding a Delete Method to the Service

1. Finally, we are on to delete. This code will look slightly different as we are introducing some powerful features available to you from the driver, builders, and filters.

2. Add the following code to your service class:

```
public void DeleteOne(string id)
{
    FilterDefinition<Game> filter = Builders<Game>
    .Filter.Eq("_id", id);
        _games.DeleteOne(filter);
}
```

We could have used the same pattern as we used previously for delete, by calling *DeleteOne* and passing in the id of the document to delete. But when doing more complex queries, builders and filters are really powerful, so this seemed a great time to introduce them.

Builders is a static helper class offering a variety of static methods to help build up queries, including filters, projections, and sorts. In the preceding code, we use *.Eq* inside the *FilterDefinition* builder to do a match against the _id field in the document and the id passed to it. But there are many other filters available, including greater than or equal to, less than or equal to, geo near for GeoJSON data, and more.

Creating a Controller

We have our service available that talks to Atlas, but as we are using controllers to define routes for our API, we need to create the controller for our Games and add the new endpoints that call the service class.

1. Create a new folder called *Controllers*.

2. Inside the folder, add a new class called *GamesController.cs*.

If you are following along using Visual Studio 2022, you can skip steps 21 to 23 by using the *API Controller – Empty* template when creating a new class.

---

3. Replace any existing using statements with the following, updating to match your project name if it differs:

```
using beginning_mongodb_atlas_dotnet.Models;
using beginning_mongodb_atlas_dotnet.Services;
using Microsoft.AspNetCore.Mvc;
```

4. Add the following attributes above your class definition:

```
[Route("api/[controller]")]
    [ApiController]
```

5. Update the class to extend from *ControllerBase*. `public class GamesController : ControllerBase` is how it should look now.

6. Create a variable for an instance of our GamesDatabaseService:

```
private readonly GamesDatabaseService _gameService;
```

7. Create a constructor that takes a GamesDatabaseService object as a parameter:

```
public GamesController(GamesDatabaseService gamesService)
        {
            _gameService = gamesService;
        }
```

This gamesService object will be available automatically because we added a singleton instance in *Program.cs* earlier.

Now we want to start adding our endpoints for each CRUD operation we added in our service class.

8. Add the following code to add a route for getting one game document that uses the *HttpGet* attribute:

```
[HttpGet]
        public ActionResult<List<Game>> Get() =>
            _gameService.Get();
```

9. Add the following code, which sets some rules on the id length and gives the route a name to differentiate it from the previous method with the same name:

```
[HttpGet("{id:length(24)}", Name = "GetGame")]
        public ActionResult<Game> Get(string id)
        {
            var game = _gameService.GetOne(id);

            if (game == null)
            {
                return NotFound();
            }

            return game;
        }
```

This one is for fetching a single game by id. It also has some error handling in the event a game document doesn't exist with that id.

We now want to move on to adding endpoints for our create methods.

10. Add the following code for inserting a single game document:

```
[HttpPost("CreateOne")]
        public ActionResult<Game> Create(Game game)
        {
            _gameService.CreateOne(game);

            return Ok();
        }
```

11. Add the following code below that for inserting many game documents at once:

```
[HttpPost("CreateMany")]
        public ActionResult<Game> CreateMany(Game[] games)
        {
```

```
        _gameService.CreateMany(games);

        return Ok();
    }
```

Two more to go. Next is update.

12. Add the following code for update:

```
[HttpPut()]
        public IActionResult Update([FromQuery]string id,
        Game gameIn)
        {
            var game = _gameService.GetOne(id);

            if (game == null)
            {
                return NotFound();
            }

            _gameService.UpdateOne(id, gameIn);

            return NoContent();
        }
```

This does similar things to previous code. Now we can move on to our final route, delete.

13. Add the following code for the delete endpoint:

```
[HttpDelete("{id:length(24)}"),]
        public IActionResult DeleteOne(string id)
        {
            var game = _gameService.GetOne(id);

            if (game == null)
            {
                return NotFound();
            }
```

```
                _gameService.DeleteOne(game.Id);

            return NoContent();
        }
```

Final Code

We have added some new classes and edited existing ones, so here is the code for each file, as it should be now.

Program.cs

```csharp
using BeginningMongoDBAtlasDotNet.Services;

var builder = WebApplication.CreateBuilder(args);

// Fetch our connection string from appsettings which is available in the
builder's Configuration
var connectionString = builder.Configuration["GamesDatabaseSettings:Connect
ionString"];

// Add services to the container.
builder.Services.AddSingleton(new GamesDatabaseService(connectionString));

builder.Services.AddControllers();
// Learn more about configuring Swagger/OpenAPI at
https://aka.ms/aspnetcore/swashbuckle
builder.Services.AddEndpointsApiExplorer();
builder.Services.AddSwaggerGen();

var app = builder.Build();

// Configure the HTTP request pipeline.
if (app.Environment.IsDevelopment())
{
    app.UseSwagger();
    app.UseSwaggerUI();
}
```

```
app.UseHttpsRedirection();

app.UseAuthorization();

app.MapControllers();

app.Run();
```

Game.cs

This class lives inside a folder called Models.

```
using MongoDB.Bson.Serialization.Attributes;
using MongoDB.Bson;
using System.ComponentModel.DataAnnotations;

namespace BeginningMongoDBAtlasDotNet.Models
{
    public class Game
    {
        [BsonId]
        [BsonRepresentation(BsonType.ObjectId)]
        public string? Id { get; set; } = String.Empty;

        [Required]
        public string Name { get; set; }

        [Required]

        public decimal Price { get; set; }

        [Required]
        public string Category { get; set; }

    }
}
```

GamesDatabaseService.cs

This class should exist inside a new folder called Services.

```csharp
using BeginningMongoDBAtlasDotNet.Models;
using MongoDB.Driver;

namespace BeginningMongoDBAtlasDotNet.Services
{
    public class GamesDatabaseService
    {
        private readonly IMongoCollection<Game> _games;
        public GamesDatabaseService(string connectionString)
        {
            var mongoDBClient = new MongoClient(connectionString);
            var database = mongoDBClient.GetDatabase("GamesDB");

            _games = database.GetCollection<Game>("Games");
        }

        public void CreateOne(Game game) => _games.InsertOne(game);

        public void CreateMany(Game[] games) => _games.InsertMany(games);

        public List<Game> Get() => _games.Find(game => true).ToList();

        public Game GetOne(string id) => _games.Find(game =>
        game.Id == id).FirstOrDefault();

        public void UpdateOne(string id, Game updatedGame) => _games
        .ReplaceOne(game => game.Id == id, updatedGame);

        public void DeleteOne(string id)
        {
            FilterDefinition<Game> filter = Builders<Game>.Filter.Eq
            ("_id", id);

            _games.DeleteOne(filter);
        }
    }
}
```

GamesController.cs

This is the final file we changed. It should exist inside a folder called Controllers.

```csharp
using BeginningMongoDBAtlasDotNet.Models;
using BeginningMongoDBAtlasDotNet.Services;
using Microsoft.AspNetCore.Mvc;

namespace BeginningMongoDBAtlasDotNet.Controllers
{
    [Route("[controller]")]
    [ApiController]
    public class GamesController : ControllerBase
    {
        private readonly GamesDatabaseService _gameService;

        public GamesController(GamesDatabaseService gamesService)
        {
            _gameService = gamesService;
        }

        [HttpGet]
        public ActionResult<List<Game>> Get() =>
            _gameService.Get();

        [HttpGet("{id:length(24)}", Name = "GetGame")]
        public ActionResult<Game> Get(string id)
        {
            var game = _gameService.GetOne(id);

            if (game == null)
            {
                return NotFound();
            }

            return game;
        }

        [HttpPost("CreateOne")]
        public ActionResult<Game> Create(Game game)
```

```
    {
        _gameService.CreateOne(game);

        return Ok();
    }

    [HttpPost("CreateMany")]
    public ActionResult<Game> CreateMany(Game[] games)
    {
        _gameService.CreateMany(games);

        return Ok();
    }

    [HttpPut()]
    public IActionResult Update([FromQuery]string id, Game gameIn)
    {
        var game = _gameService.GetOne(id);

        if (game == null)
        {
            return NotFound();
        }

        _gameService.UpdateOne(id, gameIn);

        return NoContent();
    }

    [HttpDelete("{id:length(24)}"),]
    public IActionResult DeleteOne(string id)
    {
        var game = _gameService.GetOne(id);

        if (game == null)
        {
            return NotFound();
        }

        _gameService.DeleteOne(game.Id);
```

```
        return NoContent();
    }
  }
}
```

Summary

Just like that, we have our application set up with the CRUD operation endpoints, calling into a service that interacts with Atlas.

- The MongoDB driver for C# makes it really easy to interact with your Atlas cluster.

- You can use LINQ queries with the driver.

- The operations can be achieved with single lines of code.

- There is a static helper class available called *Builder* that helps build up more complex queries.

We have our code in place but we are yet to test it, which is what Chapter 9 is all about.

CHAPTER 9

Testing the Endpoints

Now that we have endpoints for performing the create, read, update, and delete operations, it is time to test them.

ASP.NET Core Web API applications come with Swagger built in, which opens when you run the application. This allows you to see the endpoints and try them out and even comes with some nice extra features. Figure 9-1 shows an example of the Swagger page for the application we made in the previous chapters.

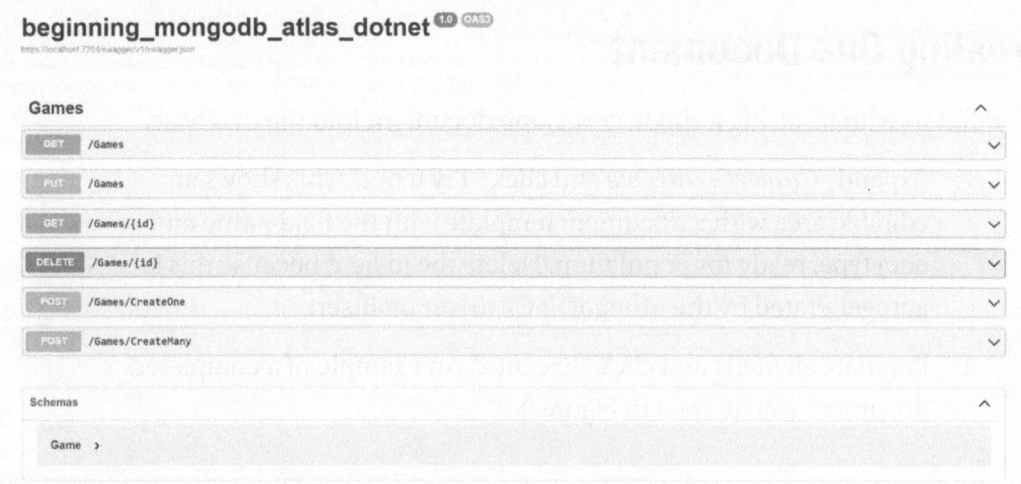

Figure 9-1. *Swagger showing the endpoints available*

It is common when working with RESTful endpoints to use clients like Postman or Insomnia to make API calls. So if you have experience with these and would prefer to use them, then you can. However, in this chapter, we are going to use Swagger as it is built in.

© Luce Carter 2024
L. Carter, *Beginning MongoDB Atlas with .NET*, https://doi.org/10.1007/978-1-4842-9550-2_9

In the event that Swagger doesn't open for you, it can be found at
https://localhost:<port your application is running on>/swagger/index.html. The
default address, if not changed, is https://localhost:7264/swagger/index.html.

Creating Documents

We don't have any data in our games collection yet, so we are going to start by creating
documents. This will also allow us to test that we can create data successfully with our
new endpoints. This uses the */Game/CreateOne* and */Game/CreateMany* endpoints,
passing a body containing the document or documents we want to insert. Swagger
makes this simple.

Creating One Document

We will start with inserting a single new game document into the database.

1. Expand */Game/CreateOne* and click "Try it out." This shows an
   editable area with a document template with the field name and
   data type, ready for population. Delete the id field because this is
   autogenerated by the MongoDB C# driver on insert.

2. Populate all fields and click "Execute." An example of a completed
   document can be seen in Figure 9-2.

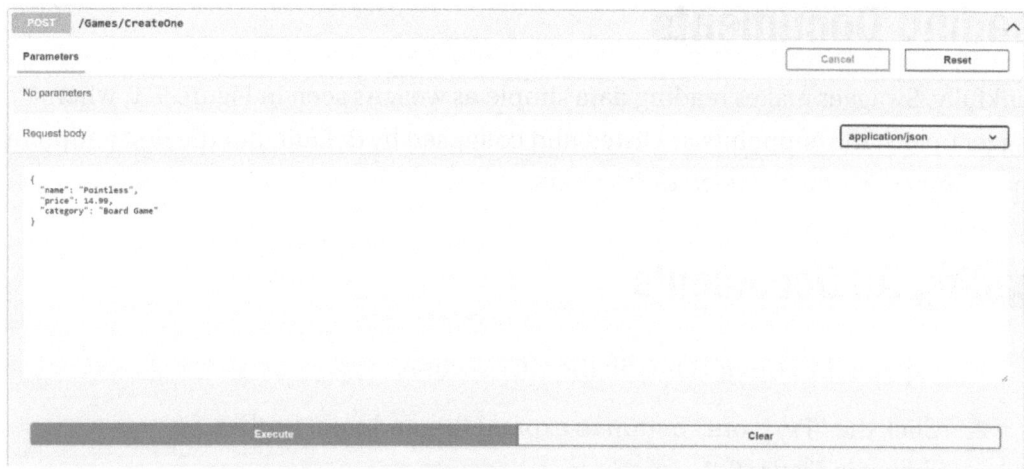

Figure 9-2. *Example document showing in Swagger*

Creating Many Documents

Testing the */Game/CreateMany* endpoint is very similar to creating one. However, instead of a template document being shown in Swagger when you hit "Try it out," you will instead see a single document inside an array.

To test this endpoint, all we need to do is populate it with multiple documents. Figure 9-3 shows an example of the array populated with multiple documents for insert.

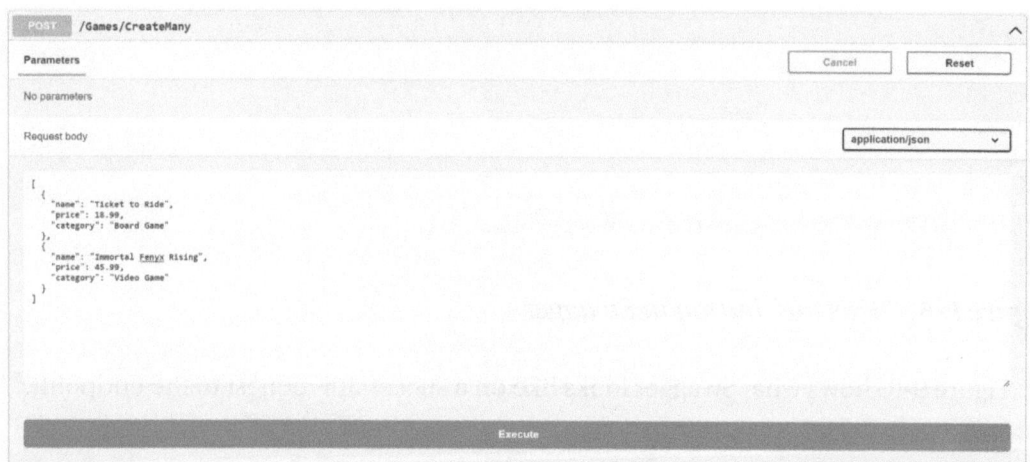

Figure 9-3. *Inserting multiple documents in Swagger*

Reading Documents

Thankfully, Swagger makes reading data simple as well. As seen in Figure 9-1, when Swagger opens, the endpoints are listed and collapsed by default. But the first endpoint listed, */Games*, is what we need to read data.

Reading All Documents

1. Expand the */Games* entry in the GET section.

2. Click the "Try it out" button to expand this endpoint further, as shown in Figure 9-4.

Figure 9-4. *"Try it out" button in Swagger*

3. This makes a blue "Execute" button appear, as shown in Figure 9-5. Click this button, which will send the *GET* request to */Games* and return back all the games you have added so far.

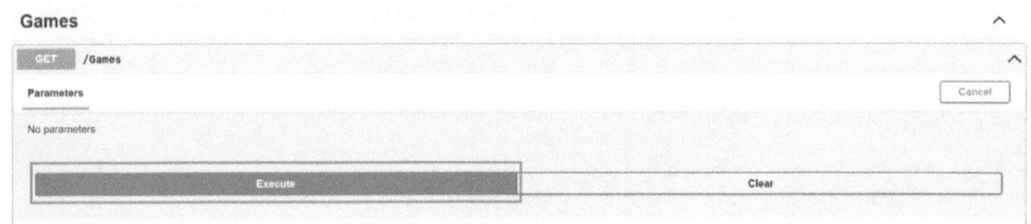

Figure 9-5. *"Execute" button in Swagger*

Figure 9-6 shows what Swagger looks like on a successful request to the endpoint.

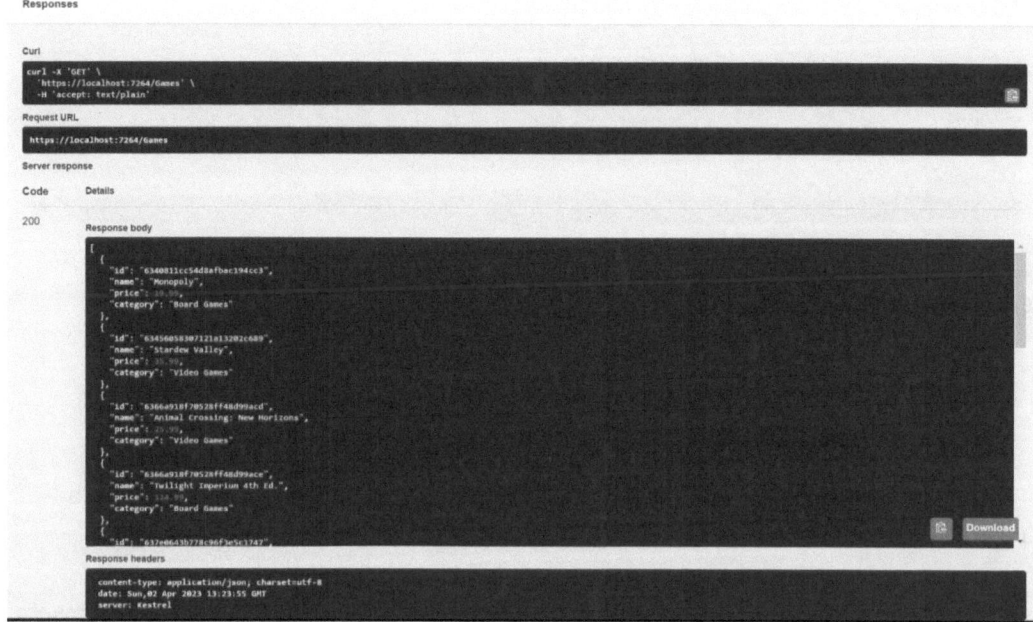

Figure 9-6. *All documents returned from endpoint in Swagger*

As shown in Figure 9-6, a successful response gives the code 200, which is the HTTP status code for OK. In most scenarios, status code 200 is what we want to see.

Reading One Document

It is time to test the */Games/{id}* endpoint for retrieving a single document. For this, we will need the id of a document that exists in the collection. You can find this by viewing the collection from any of the supported tools mentioned in earlier chapters, or by calling the */Games* endpoint to retrieve all documents and copying the id of a document that is returned.

1. Expand the /Games/{id} endpoint and click "Try it out."

2. This requires an id value to be passed, so enter it. An example is shown in Figure 9-7.

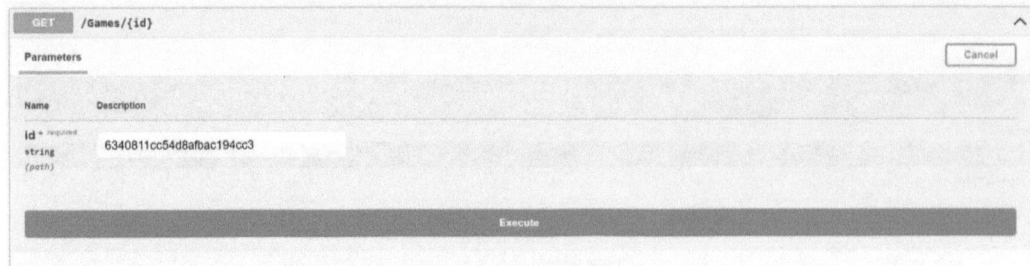

Figure 9-7. *Requesting document by id in Swagger*

3.   Click "Execute" and it will return a response similar to Figure 9-6 but for the single document you requested, if passed a valid id.

If you were to submit an id for a document that doesn't exist, you will receive a 404 status code which means "not found." An example is shown in Figure 9-8.

Figure 9-8. *Status code 404 error when requesting id that does not exist*

Updating a Document

What happens if you spot an incorrect piece of information in a document or you want to update something because it has changed? This is where update is useful, and we can take advantage of the */Games* endpoint available in Swagger by using the PUT method.

1.   Testing this endpoint requires knowing the id of the document you wish to update. So use the endpoint for reading all the documents to find the id of a single document you want to update.

2.   Enter the id for the document you wish to update in the id box.

3. Enter the new details for the document in the editable area, ensuring you populate the id field with the same value. An example is shown in Figure 9-9.

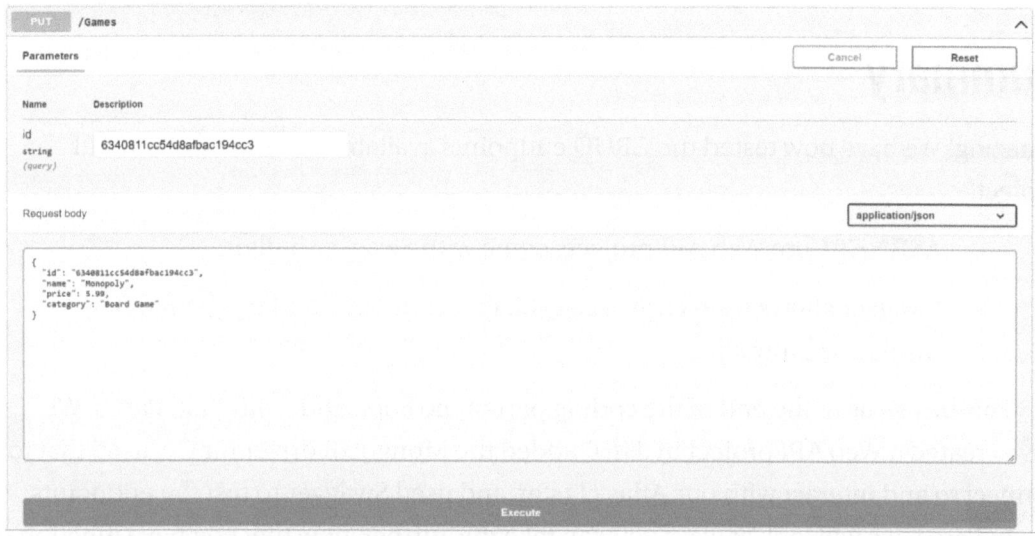

Figure 9-9. *Updating existing document in Swagger*

4. Click "Execute," and this will test that update endpoint.

5. If you wish, you can run the */Games/{id}* endpoint again with the id of the document you just updated to see if it has changed.

Deleting a Document

Finally, for the last section of this chapter, we will test the delete endpoint.

1. Testing this endpoint also requires knowing the id of the document you wish to delete. So, as before, by using the endpoint for reading all documents, you can find the id of the document to delete.

2. Enter the id of the document to delete in the id box, as seen in Figure 9-9.

3. Press "Execute" to request the deletion of the document by that id.

If it successfully finds a document with that id, it will be deleted from the database. You can always call /Games{id}, passing the id of the document you just deleted to see if you get a 404 – a "not found" response – showing it has been successfully deleted.

Summary

Amazing! We have now tested the CRUD endpoints available to us in our Web API project.

- ASP.NET Core Web API projects come with Swagger built in.
- Swagger allows us to easily see available endpoints and try them out inside our browser.

This brings us to the end of the coding part of the book and, with that, Part 3. We have created a Web API project in .NET, added the MongoDB driver for C#, used that to connect to and interact with our Atlas cluster, and used Swagger to test the endpoints.

In Part 4, we will look at ways you can take this further, now that you have functional code that uses architecture that is useful for more complex applications. In the next chapter, we will look at schema design, for those times when you want to specify the shape of your documents.

PART IV

Taking It Further

CHAPTER 10

Schema Validation

As you learned in earlier chapters, one of the advantages of MongoDB is its flexible schema, meaning that not all documents have to have the same shape. The term *shape* refers to the fields that are present. This means that you could add a document that has only some or even none of the fields present in another document.

However, this isn't always ideal in production, and many people find that they want to enforce the document shape. This can include the fields, value limitations such as character length, and even data type.

In this chapter, you will learn how MongoDB allows you to set up and take advantage of schema validation, by seeing how you might apply some validation to our games database.

Schemas can be defined and rules set from within many of the MongoDB tools, including Compass, Atlas, and the tool shown in this chapter, *mongosh*.

Data Modeling

Before we go into schema validation, I want to take a few paragraphs to touch on data modeling as this has an impact on your schemas.

This is because how you model your data plays a key role in so much of your application. If you don't at least think about this before you start, you won't necessarily model your data in a way that is future-proof and leads to good performance should you be storing, querying, and manipulating large amounts of data. Don't get me wrong, MongoDB can function extremely well out of the box, but it is still good practice to consider this.

© Luce Carter 2024
L. Carter, *Beginning MongoDB Atlas with .NET*, https://doi.org/10.1007/978-1-4842-9550-2_10

For starters, if you come from a relational database background, you might be used to normalizing your data, but in fact, with MongoDB, you are better off leaving your data denormalized. This is because it focuses on embedding over spreading related data across collections. You will often hear a mantra from MongoDB staff that we think is so true: *"Data that is accessed together should be stored together."* If the data is related, you will get far better performance keeping it together in a single document.

There are, of course, reasons to form relationships across collections – for example, a customer collection and an orders collection. You might embed a list of order ids inside a customer document to show what they have bought, but the larger details of the order might be stored in an order collection document where the *_id* of that document is the *order_id* in the customer document.

How you model your data comes down to how it will be used by your application. Embedding is always preferred. Although it is out of the scope of this book, there are many resources available on data modeling with MongoDB, including written articles and webinars, if you wish to learn more.

Validating for Required Fields

Even though your documents can have whatever fields that you want, you might have some that you consider vital that you want to ensure appear in every document. This is where required fields come in.

The following code example shows how you might add validation to ensure that the name, price, and category fields in the games document are present:

```
db.runCommand(
    {
        collMod: "Games",
        validator: {
            $jsonSchema: {
                bsonType: "object",
                required: [ "name", "price", "category" ]
            }
        }
    }
);
```

The preceding command is slightly different from what you would run if the collection didn't already exist.

In this instance, we are running the command collMod which means collection modification. We then pass the validator field a JSON schema specifying our required fields. You will notice that everything in MongoDB, even schemas, uses documents. So, our JSON schema is a document of fields the validator can understand, specifying what we want.

The following code is what this validation would look like if applied during collection creation:

```
db.createCollection("Games",
    {
        validator: {
            $jsonSchema: {
                bsonType: "object",
                required: [ "name", "price", "category" ]
            }
        }
    }
);
```

This is very similar to the first command. But this means that should you wish to apply validation from the start, you can.

Figure 10-1 shows what the error would look like if you tried to add a document to the collection that doesn't meet the validation rules.

```
Atlas atlas-9ljt0v-shard-0 [primary] GamesDB> db.Games.insertOne({"price": "9.99"});
Uncaught:
MongoServerError: Document failed validation
Additional information: {
  failingDocumentId: ObjectId('65b65b2574a52e53d0129ec5'),
  details: {
    operatorName: '$jsonSchema',
    schemaRulesNotSatisfied: [
      {
        operatorName: 'required',
        specifiedAs: { required: [ 'name', 'price', 'category' ] },
        missingProperties: [ 'category', 'name' ]
      }
    ]
  }
}
```

Figure 10-1. *Validation failure on required fields*

The great thing is that when the validation failed, without needing to specify any extra details, the server automatically returned an error explaining why it failed and which fields were missing.

Handling Invalid Documents

However, you might not always want it to simply fail if you attempt to insert or modify a document that doesn't meet the validation rules you set.As you saw from Figure 10-1, the default behavior is to reject any updates or creations that fail that validation.

If you would rather it didn't error but simply recorded an error in the log – this is also possible.

As well as passing the validator containing the JSON schema, you can also pass a validationAction property of "warn" instead of the default value of "error."

This is what the earlier code example for your existing collection would look like if you passed a warn value instead:

```
db.runCommand(
    {
        collMod: "Games",
        validator: {
            $jsonSchema: {
                bsonType: "object",
                required: [ "name", "price", "category" ]
            }
        },
        validationAction: "warn"
    }
);
```

As you can see, it is mostly identical, except for the additional property. If you attempted to add a document now that didn't pass validation, it would still get created, but if you viewed the log, you would see an entry like Figure 10-2.

{"t":{"$date":"2024-01-28T14:28:55.730+00:00"},"s":"W", "c":"STORAGE", "id":20294, "ctx":"conn34736","msg":"Document would fai
l validation","attr":{"namespace":"GamesDB.Games","document":{"_id":{"$oid":"65b664a774a52e53d0129ec8"},"price":"9.99"},"errInfo":
{"failingDocumentId":{"$oid":"65b664a774a52e53d0129ec8"},"details":{"operatorName":"$jsonSchema","schemaRulesNotSatisfied":[{"oper
atorName":"required","specifiedAs":{"required":["name","price","category"]},"missingProperties":["category","name"]}]}}}}

Figure 10-2. *Document that would fail validation showing in the logs*

Specifying Data Types for Fields

So far, you have seen how to specify what fields must be present in the document. But what about if you want to apply rules to what type of value is present in a field – for example, insisting that the name field must be a string? This is also possible and straightforward to achieve.

The following code example shows the same validation as you set up earlier for required fields but also passes a set of documents containing the fields for which to set a type:

```
db.runCommand(
    {
        collMod: "Games",
        validator: {
            $jsonSchema: {
                bsonType: "object",
                required: [ "name", "price", "category" ],
                properties: {
                    "name": {
                        bsonType: "string",
                        description: "must be a string and is required"
                    },
                    "price": {
                        bsonType: "double",
                        description: "must be a double and is required"
                    }
                }
            }
        },
        validationAction: "error"
    }
);
```

The validationAction field has been added here because it is updating an existing schema where the validation had previously been set to warn. But if this is still the default error value, it wouldn't be required.

Figure 10-3 shows attempting to add a document with the wrong data type for price and the resulting error, showing the description defined in the schema.

```
Atlas atlas-9ljt0v-shard-0 [primary] GamesDB> db.Games.insertOne({"name": "SushiGo", "price": "9.99", "category": "Card Games"});
Uncaught:
MongoServerError: Document failed validation
Additional information: {
  failingDocumentId: ObjectId('65b66bba74a52e53d0129ecd'),
  details: {
    operatorName: '$jsonSchema',
    schemaRulesNotSatisfied: [
      {
        operatorName: 'properties',
        propertiesNotSatisfied: [
          {
            propertyName: 'price',
            description: 'must be a double and is required',
            details: [
              {
                operatorName: 'bsonType',
                specifiedAs: { bsonType: 'double' },
                reason: 'type did not match',
                consideredValue: '9.99',
                consideredType: 'string'
              }
            ]
          }
        ]
      }
    ]
  }
}
```

Figure 10-3. *A document failing data type validation and the resulting error*

What is clever as well is not only is the description returned, but the server can also provide details on what it thinks you attempted to use instead – in this case, a string instead of a double.

Specifying Allowed Values for a Field

Another useful type of validation is limiting the values of a field to only certain ones. This might be helpful in the games document, for example, to offer a list of categories for the category field, such as board game, card game, and video game.

This uses the same properties field you saw in the data type example, but this time, you can provide a list of choices. The following code example shows a validation schema limiting the values for the category field:

```
db.runCommand(
    {
        collMod: "Games",
        validator: {
            $jsonSchema: {
                bsonType: "object",
                required: [ "name", "price", "category" ],
```

```
        properties: {
            "category": {
                enum: ["Board Game", "Video Game", "Card Game"],
                description: "Must be either Board Game, Video
                Game, or Card Game"
            }
        }
    }
}
);
```

Figure 10-4 shows what happens if you attempt to add a document with a category value outside the specified list.

```
Atlas atlas-9ljt0v-shard-0 [primary] GamesDB> db.Games.insertOne({"name": "Moss: Book 2", "price": 24.99, "category": "VR Game"});
Uncaught:
MongoServerError: Document failed validation
Additional information: {
  failingDocumentId: ObjectId('65b66f6c74a52e53d0129ed2'),
  details: {
    operatorName: '$jsonSchema',
    schemaRulesNotSatisfied: [
      {
        operatorName: 'properties',
        propertiesNotSatisfied: [
          {
            propertyName: 'category',
            description: 'Must be either Board Game, Video Game, or Card Game',
            details: [
              {
                operatorName: 'enum',
                specifiedAs: { enum: [ 'Board Game', 'Video Game', 'Card Game' ] },
                reason: 'value was not found in enum',
                consideredValue: 'VR Game'
              }
            ]
          }
        ]
      }
    ]
  }
}
```

Figure 10-4. *Attempting to use a value for the category field not in the specified list*

Again, the server does a great job of explaining not only that it failed validation because it wasn't one of the options for category but also clearly shows what the possible valid options are.

Applying Validation to Existing Documents

You may have noticed that so far, all validation has been applied to and tested against new documents. This is because if you are applying a schema to an existing collection like the games collection, it doesn't automatically validate existing documents.

This is set using a property called validationLevel which can be either "*strict*" or "*moderate.*"

Strict is the default, and this means that if you attempt to update an existing document that fails your new schema, the update will fail.

The following code example shows passing validation that sets the validationLevel to strict:

```
db.runCommand(
    {
        collMod: "Games",
        validator: {
            $jsonSchema: {
                bsonType: "object",
                required: [ "name", "price", "category" ],
                properties: {
                    "price": {
                        bsonType: "string",
                        description: "Must be a string and is required"
                    }
                }
            }
        },
        validationLevel: "strict"
    }
);
```

If an existing document is updated now and something about it doesn't match the schema, it will error, as shown in Figure 10-5.

```
Atlas atlas-9ljt0v-shard-0 [primary] GamesDB> db.Games.updateOne(
...    { _id: ObjectId("65b6697e74a52e53d0129eca") },
...    {
...      $set: {
...        name: "Sushi Go",
...        Price: 24.99,
...        Category: "Card Game"
...      }
...    }
... );
Uncaught:
MongoServerError: Document failed validation
Additional information: {
  failingDocumentId: ObjectId('65b6697e74a52e53d0129eca'),
  details: {
    operatorName: '$jsonSchema',
    schemaRulesNotSatisfied: [
      {
        operatorName: 'properties',
        propertiesNotSatisfied: [
          {
            propertyName: 'price',
            description: 'Must be a string and is required',
            details: [
              {
                operatorName: 'bsonType',
                specifiedAs: { bsonType: 'string' },
                reason: 'type did not match',
                consideredValue: 9.99,
                consideredType: 'double'
              }
            ]
          }
        ]
      }
    ]
  }
}
```

Figure 10-5. *Error when updating an existing document that doesn't match new schema*

The schema had been updated to set the data type for price to a string and strict validation, so when the document was updated with price as a double, the server knew and returned an error.

The other option is to set validationLevel to moderate which means that if existing documents fail validation, then they can still be updated. This is useful if you don't necessarily care if a document added prior to the schema passes validation or not.

Allowing Invalid Documents on an Ad Hoc Basis

Sometimes, you might find that you want to allow a document to be inserted or updated, even if it fails validation. For these instances, there is a property to allow this, bypassDocumentValidation.

If you set this to true as part of the command, it will allow it to happen.

```
db.runCommand( {
    insert: "Games",
    documents: [
        {
            name: "Fluxx",
            price: 6.99,
            category: "Card Game"
        }
    ],
    bypassDocumentValidation: true
} );
```

The price field fails validation as it is being passed a decimal and not a string, but because the field is set to true, it will be allowed.

Finding Invalid Documents to Update

Instead of waiting for a document update to occur to highlight documents that don't pass schema validation, you may choose to manually identify any that will require updating.

The first step is to define the schema as a variable that can be used later:

```
let myschema =
{
    $jsonSchema: {
        required: [ "name", "price", "category" ],
```

```
    properties: {
        name: { bsonType: "string" },
        price: { bsonType: "string" },
        category: {
            enum: ["Board Game", "Card Game", "Video Game"],
            description: "Must be either Board Game, Card Game, or
            Video Game"
        }
    }
  }
}
```

Then, you can run a find query against your collection to find any that do not match that schema:

```
db.Games.find({$nor: [myschema]});
```

The $nor here means "not or" and is asking the database to find any documents that do not match the query – so, our schema. By performing this query, passing the schema defined earlier, it will return any documents that do not meet the schema definition.

Figure 10-6 shows the output of searching the games database and finding documents that do not meet the validation rules.

```
Atlas atlas-9ljt0v-shard-0 [primary] GamesDB> db.Games.find({$nor: [myschema]});
[
  {
    _id: ObjectId('658ad04ae82fc9458f658265'),
    Name: 'Monopoly',
    Price: '19.99',
    Category: 'Board Games'
  },
  { _id: ObjectId('65b660b474a52e53d0129ec6'), price: '9.99' },
  { _id: ObjectId('65b6627074a52e53d0129ec7'), price: '9.99' },
  { _id: ObjectId('65b664a774a52e53d0129ec8'), price: '9.99' },
  { _id: ObjectId('65b664dc74a52e53d0129ec9'), price: '9.99' },
  {
    _id: ObjectId('65b6697e74a52e53d0129eca'),
    name: 'SushiGo',
    price: 9.99,
    category: 'Card Games'
  },
  {
    _id: ObjectId('65b669d174a52e53d0129ecb'),
    name: 'SushiGo',
    price: '9.99',
    category: 'Card Games'
  },
  {
    _id: ObjectId('65b6796774a52e53d0129ed4'),
    name: 'Spiderman 2',
    price: 59.99,
    category: 'Video Game'
  }
]
```

Figure 10-6. *An array of documents from the collection that don't meet the new schema*

If you wanted, you could combine an updateMany operation with a $set stage to not only find invalid documents but also update them, perhaps with a field stating the document is not valid:

```
db.Games.updateMany(
    {
        $nor: [ myschema ]
    },
    {
        $set: { isValid: false }
    }
);
```

Running the preceding command would result in the output shown in Figure 10-7, showing that the new field was added to documents in the collection.

```
Atlas atlas-9ljt0v-shard-0 [primary] GamesDB> db.Games.updateMany( { $nor: [myschema] }, { $set: { isValid: false } } );
{
  acknowledged: true,
  insertedId: null,
  matchedCount: 8,
  modifiedCount: 8,
  upsertedCount: 0
}
```

Figure 10-7. *Server showing that invalid documents were updated with an isValid field*

Schema Anti-Patterns

You have seen how you can apply different types of schemas to your application and handle validating against them. But creating the perfect schema the first time isn't always possible. To make it easier, there is a feature inside Atlas that can help identify areas for improvement.

When you are in the data explorer in Atlas viewing a collection, there is a tab along the top called *Schema Anti-Patterns*, as shown in Figure 10-8. If there are any common problems in your schema that could be improved, this will identify them and show you how to improve, including links to the relevant documentation to help you learn more. Handy!

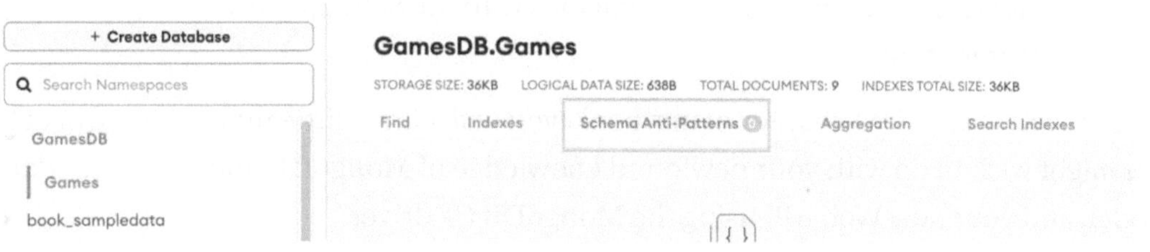

Figure 10-8. *Schema Anti-Patterns tab showing in Atlas*

The games collection only has a small number of documents with very few fields inside, so there are no anti-patterns identified. But they would show up in this section if there were.

An example of an anti-pattern might be having too many sub-documents in an array within the document, which can impact performance. The solution to this is often to move the documents to a separate collection and then reference them from the original document.

But remember what was said earlier – *"Data that is accessed together should be stored together"* – so referencing should never be the default but instead something you do when you have a good reason. Referencing may feel natural if you come from a relational database system where referencing via keys between tables is normal, but to enjoy the power of the document database model, it does require a mind shift away from spreading data across "tables," a.k.a. collections, from the beginning.

Summary

Data modeling and schemas are a very important part of making the most of databases. This is just as true for MongoDB's document database model. In this chapter, you have learned the following:

- MongoDB's flexible schema doesn't mean you can't impose a schema on your collection.

- You can apply rules on that schema to specify how to handle invalid documents.

- The different types of schemas and validation you can apply, including data types, required fields, and restricting possible values.

- How to identify invalid documents.

- How Atlas can help improve schemas with the Schema Anti-Pattern tool.

In the next and final chapter of this book, you will learn a little more about what else you might wish to do with your newfound knowledge of MongoDB, the document data model, and your new Web API using the MongoDB C# driver.

CHAPTER 11

What Next?

Welcome to the final chapter of *Beginning MongoDB Atlas with .NET*. What a journey!

At this point, you now know all about MongoDB, the document database model, and the developer data platform suite of products known as Atlas, and you have a working .NET Web API project that uses the MongoDB C# driver to talk to Atlas.

In this chapter, you will learn about what you might want to do next, now that you have this new knowledge.

Adding a UI

Currently, you just have an API you can call endpoints of to perform the CRUD operations on your data. But you could also introduce a UI to display the information visually to an end user.

There are a few options for this:

1. Adding a new project to the solution using one of the ASP. NET Core UI frameworks such as *Blazor* or *MVC (Model-View-Controller)* that calls the API

2. Creating a brand-new front-end project and using the MongoDB C# driver from within that project

3. Using *.NET MAUI* and Atlas Device SDKs to create a cross-platform mobile and desktop application; you could even combine it in a *Blazor Hybrid* app to also create a mobile app from existing website code

How to achieve these is outside of the scope of this book, so it won't be covered in detail here, but option three is slightly different from the first two options.

© Luce Carter 2024
L. Carter, *Beginning MongoDB Atlas with .NET*, https://doi.org/10.1007/978-1-4842-9550-2_11

Using .NET MAUI and Atlas Device SDKs

If you are using .NET MAUI (MAUI) to create a cross-platform mobile and desktop app, you won't use the MongoDB C# driver as used in the rest of the book.

This is because not only is a MAUI project structured differently to the ASP.NET Core family, such as Blazor and Web API projects, but the requirements are different and use a different driver: Atlas Device SDKs.

As you learned in Chapter 2, Atlas Device SDKs are great because they allow you to overcome the complexity of Internet connectivity that comes with a mobile app. It does this by providing flexible sync which automatically keeps a local database stored on the device in sync with Atlas, meaning any app will continue to interact with the database, even when offline.

But the great thing is that even though Atlas Device SDKs are a different driver to the MongoDB C# driver, they can still point to the same Atlas cluster so you can have multiple apps and websites that use the same data.

So, you might choose to learn more about .NET MAUI and Atlas Device SDKs and create an app that talks to the same database you used in your Web API project.

Visualizing Your Data

Another fantastic product covered in Chapter 2 is Atlas Charts. This allows you to visualize your data without needing any third-party graphing libraries or writing any code yourself.

You can create a variety of chart types and base it on all kinds of data. This is because not only will it see the raw data stored in your collections, but you can carry out aggregations and pre-processing on the data that doesn't impact the database but gives you more power over what is shown in the charts.

For example, you could run an aggregation to calculate the average price of the games in your games collection and create a line graph that shows how the average price you spend changes over time.

You could even get more granular by adding a new date field to your documents, called createdAt or similar, that helps you track data changes or calculations over time.

There is also a new feature now within tools such as Charts and Compass that will use GenAI to help generate queries and aggregations, so if you know what you want to achieve but not how, you can use that to help you!

Within Charts, you make a dashboard to hold one or more charts, and it will even provide code for embedding your dashboard into your site.

If you add a UI such as a Blazor front end, give it a go and add a Charts dashboard.

Searching Your Data

Atlas provides another two great products you might want to implement: Search and Vector Search.

Atlas Search

Search is a full-text search that allows users to search by keywords. So, if your games collection or any other is particularly large, you might wish to provide the ability to search on keywords against fields in the data such as name.

Inside Atlas, you can add a search index to your data and even specify a field mapping to be of type "autocomplete" which means you can support autocomplete.

The MongoDB C# driver even supports Atlas Search so you can run the queries needed to search against the data from within your application.

Plus, you can add fuzzy search options, allowing some flexibility around exact matches and spelling mistakes.

Atlas Vector Search

Sometimes, searching by keyword isn't enough. You might want to search by meaning or among unstructured data such as images or videos. This is where vector search comes in. This is another feature supported by an Atlas product of the same name.

You could use a third-party service like OpenAI to generate embeddings for your collection, add them as a new field, and take advantage of the expanded search benefits of vector searching.

Play with the Sample Data

The games database alone may not be enough data at this stage to fully take advantage of some of the products discussed in Chapter 2, including the ones mentioned earlier in this chapter.

This is where the sample datasets are fantastic. They cover a broad range of data shapes and sizes with many documents within each for a good sample size. There are even collections within the sample dataset now with embeddings for Atlas Vector Search. This can make them a fantastic playground for trying out features and products before applying them to your own scenarios.

Learn More

This book served as a starting point for introducing you to MongoDB and the C# driver, but there were plenty of topics outside the scope of the book.

A great place to visit next is MongoDB's Developer Center. This is a hub of tutorials, quick starts, articles, podcasts, videos, and more in not just .NET and C# but other languages and even language-agnostic topics, such as products, performance, and much more. The content there is written and reviewed by many MongoDB experts. Developer Center is a helpful curation tool.

The documentation is also very useful, especially if you want to understand specific commands and required parameters related to a driver.

Last but by no means least is the forums. MongoDB has a very active community forum that is frequented by employees, community champions, enthusiasts, and many more. You can search for answers to questions you have that have been answered publicly before, ask new ones, or even provide your own answers with your new knowledge of MongoDB!

Summary

With that, you have reached the end of the book. Wow!

In this chapter, you have seen a few suggestions about where you might go next:

- Add a UI to view your data from a web page.

- Use the Atlas Device SDKs with .NET MAUI to create a cross-platform mobile and desktop app that shares the same data.

- Visualize your data with Atlas Charts.

- Add the ability to search your collection either via full-text search with Atlas Search or via semantics with Vector Search.

- Take advantage of the sample datasets.

- Visit the Developer Center, documentation, or forums to learn more.

Thank you so much for taking the time to read, follow, and hopefully enjoy this book. It is only the start of your journey with MongoDB, showing you how sticking to the document database model can save you time and bring a lot of benefits, especially when combined with the power of the Atlas developer data platform.

Happy coding and happy learning!

Index

A

Adding model, 121, 122
Aggregation, 27–29, 31, 72
Aggregation builder tool, 78–81
Apache Cassandra, 13
ASP.NET Core, 108, 114
ASP.NET Core UI frameworks, 161
ASP.NET Core Web API project,
     102, 104–108
ASP.NET Core Web API, Swagger, 137
  creating documents
    /Game/CreateMany endpoints, 139
    /Game/CreateOne endpoints, 138
  CRUD endpoints, 144
  deleting document, 143, 144
  reading documents
    reading all documents, 140, 141
    reading one documents, 141, 142
  updating document, 142, 143
ASP.NET Core Web Application, 107
Atlas account, Google
  deployment, 49
  details, 45, 46
  email verification, 47, 48
  questions, 48, 49
Atlas App Services, 32, 86, 91
  authentication, 33
  data API, 34
  Serverless functions, 32, 33
  static hosting, 34
  triggers, 32
Atlas Device SDKs, 162, 165

Atlas Search, 163
Atlas UI, 61, 73
  application, 112
  collections view, 63
  connection string, 112
  dashboard, 61
  data collections, 62
  data type, 65
  deleting data, 67
  document area, 67
  documents, 64
  JSON documents, 66
  JSON view, 67
  sample_restaurants database, 64
Atlas Vector Search, 163, 164
Atomicity, 7
Atomicity consistency isolation durability
    (ACID), 7
Authentication, 33, 35, 59
Authentication providers, 33
avgRating, 79
AWS' DynamoDB, 12

B

Beginning MongoDB Atlas with .NET
  adding UPI, 161
  Atlas Device SDKs, 162
  Atlas Search, 163
  Atlas Vector Search, 163
  data visualization, 162, 163
  documentation, 164

167

Beginning MongoDB Atlas with .
NET (*cont.*)

.NET MAUI (MAUI), 162

sample data, 164

Browsing data, 83

bypassDocumentValidation, 156

C

Caching, 12

Charts, 35, 36, 162, 165

Cloud console, 35, 51, 61

collMod command, 149

Command-line interface (CLI), 31, 68, 102

Command-line interface (CLI) tools,
25, 68, 102

connectionString, 124

Connection string, 68, 69, 74, 75, 112

Consistency, 7, 25, 35

createdAt, 162

Creating and manipulating data, 77

Creating controller, 126–129

Creating data, 72, 88–89

CRUD operation endpoints, 135

D

Data, 3

Data access objects (DAOs), 34

Data API, 34, 86–89, 91

Data API endpoint, 87

Database, 3

NoSQL database (*see* NoSQL database)

relational database (*see* Relational
database)

replica sets, 24, 25

sharding, 25–27

tables, 3

Data Federation, 30

Data Lake, 31, 35

Data modeling, 147, 148, 160

deleteOne function, 85

Deleting data, 73, 77, 85, 95–97

DoubleClick, 21

Durability, 7

E

Edges, 14

Efficiency, 8, 12

Embedding, 18, 35, 148, 163, 164

Enable Data API, 87

F

findOne method, 71

G

/Game/CreateMany
endpoints, 138, 139

Game.cs, 21, 131

Game document, 125, 127, 128, 138

/Games endpoint, 141, 142

GamesController.cs, 126, 133

GamesDatabaseService.cs, 123, 124, 132

GamesDatabaseService object, 127

/Games/{id} endpoint, 141, 143

gamesService object, 127

GetWeatherForecast endpoint, 107

GitHub account, 42, 44, 50

Google account, 42–44, 50

Graphical user interface (GUI)
application, 73

GraphiQL, 95, 96

GraphQL, 91–97

H

Horizontal scaling, 19, 21, 25
HttpGet attribute, 127

I

insertMany method, 72
InsertOne method, 72, 124
Integrated development environment
    (IDE), 102, 105–108
Internet of Things (IoT), 13
IP access restriction lists, 59
Isolation, 7

J, K

JavaScript Object Notation (JSON), 10, 17
JetBrains Rider, 105–107
JSON schema, 149, 150

L

Label, 15

M

Memcached, 12
Microsoft *Azure*, 19, 24, 32
Microsoft's SQL server, 5
Models, 131
MongoDB, 20, 21, 36, 85, 90, 92, 112, 117,
    119, 164, 165
    Atlas, 22
    Atlas database and collections, 111
    Change Streams, 22
    NET, 111
    schema validation, 22
    versions, 22

MongoDB account, 41, 42
MongoDB Atlas, 19, 36, 61
    aggregation, 28, 29
    CLI, 31
    cloud console, 51
    database
        replica sets, 24, 25
        sharding, 25–27
    Data Federation, 30
    Data Lake, 31
    document database as a service, 24
    first database cluster, 51
        build database, 51, 52, 59
        Cloud Provider/region, 53
        configuration, 55
        create cluster, 53, 54
        creation, 56
        IP address to access list, 55
        Security Quickstart window, 54
        Shared cluster, 52
    sample dataset
        confirmation, 57, 58
        ellipses button, 56, 57
        loading, 57, 58
    Search, 31
    serverless, 30
MongoDB C# driver, 111, 138, 160–163
MongoDB client, 112, 115, 116, 119, 124
MongoDB Management Services
    (MMS), 22
MongoDB's Developer Center, 164
MongoDB's developer data, 101
MongoDB's document database
    model, 160
MongoDB server (on-premises), 23
MongoDB Shell, 69, 70, 73, 80, 84, 85, 96
MongoDB Shell syntax, 80
MongoDB University, 22

Mongosh, 68–72
Monitoring, 22
Multi-dimensional key-value store, 13
MySQL, 5

N

.NET development, 102
Netflix, 16
.NET MAUI (MAUI), 162
Nodes, 14, 15
NoSQL database
    document databases
        advantages, 19
        data, 18
        disadvantages, 20
        employees data, 17, 18
        features, 17
        flexibility, 18
        JSON, 17
        key-value-based fields, 18
        MongoDB, 20
        recreation, 18
        uses, 18
    graph databases, 14
        advantages, 16
        disadvantages, 16, 17
        features, 14, 15
        Netflix, 16
        nodes, 15
        train journey, 15
    key-value/key-store, 10
        advantages, 12
        Couchbase, 12
        disadvantages, 12, 13
        features, 10
        game setting, 11

Redis, 12
Windows Registry, 10, 11
    uses, 9
    wide-column databases, 13
        advantages, 13, 14
        Apache Cassandra, 13
        disadvantages, 14
        IoT, 13
        series data, 13
NuGet packages, 102, 111, 112, 115, 119

O

Object-relational mappers (ORMs), 34
Official drivers, 90
Oracle, 5
Oracle SQL, 5

P, Q

Package Manager Console, 112
Platform as a service (PaaS), 21
PostgreSQL, 5
Program.cs file, 117, 124, 127, 130

R

Reading data, 71, 74
Realm, 34, 36, 91
    Atlas database, 35
    Charts, 35, 36
    Device Sync, 35
    local databases, 34
    mobile database product, 34
    SDKs, 35
Relational database, 20, 148
    advantages

ACID, 7
  data accuracy, 8
  normalization, 6
  simplicity, 6
columns, 4
data, 4
disadvantages, 8, 9
foreign key, 4
orders table, 4
primary key, 4
product table, 4
providers, 5
queries, 5
SQL, 5
tables, 3, 5
Relational database management system
    (RDBMS), 3, 5
ReplaceOne method, 125
RESTful endpoints, 137

S

Sample dataset, 59
Scalability, 8, 12, 17, 19
Schema Anti-Patterns, 159, 160
Schemas, 147
Schema validation
  applying
    finding invalid documents, 156,
        157, 159
    invalid documents, 156
    updateing exiting document,
        154, 155
    validationLevel, 154, 156
  handling invalid documents, 150
  for required fields, 148–150
  schema Anti-Patterns, 159, 160

specifying allowed values for fields,
    152, 153
  specifying data types for fields, 151, 152
Search, 31, 163, 164
Server Side Public License (SSPL), 23, 161
Services, 123
  adding Create methods, 124, 125
  adding Delete methods, 126
  adding Read methods, 125
  adding Update methods, 125
  creation, 123
  updating to use, 124
Shape, 147
Static hosting, 34, 36
Structured query language (SQL), 5, 6, 12,
    16, 23, 27
Swagger, 121, 137–144
Swagger page, 108, 116

T

Traditional database as a service, 30
Transact-SQL (T-SQL), 5

U

updateMany operation, 95, 158
updateOne, 72, 85, 89
Updating data, 64–67, 72, 85, 89
User interface (UI), 3, 56, 64, 67, 73, 77,
    161, 165

V

validationAction property, 150
Validation rules, 149, 150, 157
Vertical scaling, 8, 21, 26

Visual Studio 2022, 105, 106, 126
Visual Studio Code (VS Code), 81–86, 102
VS Code Extension, 81, 82, 86

W

WeatherForecast.cs, 104, 108
Web API, 104

Web API project, 101–108,
114, 121, 137, 144,
160, 162

X, Y, Z

XML, 17